INDUCED EFFECTS
OF GENOTOXIC AGENTS
IN EUKARYOTIC CELLS

INDUCED EFFECTS OF GENOTOXIC AGENTS IN EUKARYOTIC CELLS

Edited by

Toby G. Rossman
NYU Medical Center
Institute of Environmental Medicine
New York, New York

● **HEMISPHERE PUBLISHING CORPORATION**
A member of the Taylor & Francis Group
Washington   Philadelphia   London

USA	Publishing Office:	Taylor & Francis
		1101 Vermont Avenue, NW, Suite 200
		Washington, DC 20005-3521
		Tel: (202) 289-2174
		Fax: (202) 289-3665
	Distribution Center:	Taylor & Francis
		1900 Frost Road, Suite 101
		Bristol, PA 19007-1598
		Tel: (215) 785-5800
		Fax: (215) 785-5515
UK		Taylor & Francis Ltd.
		4 John St.
		London WC1N 2ET
		Tel: 071 405 2237
		Fax: 071 831 2035

INDUCED EFFECTS OF GENOTOXIC AGENTS IN EUKARYOTIC CELLS

1 2 3 4 5 6 7 8 9   B R B R   9 8 7 6 5 4 3 2

This book was set in Omega by Hemisphere Publishing Corporation. The editors were Elizabeth Dugger and Lisa Speckhardt; the production supervisor was Peggy M. Rote; and the typesetter was Wayne Hutchins. Cover design by Michelle Fleitz.
Printing and binding by Braun-Brumfield Inc.

A CIP catalog record for this book is available from the British Library.
∞ The paper in this publication meets the requirements of the ANSI Standard Z39.48-1984(Permanence of Paper)

Library of Congress Cataloging-in-Publication Data

Induced effects of genotoxic agents in eukaryotic cells / edited by
  Toby G. Rossman.
      p. cm.
    Includes bibliographical references and index.

    1. Chemical mutagenesis. 2. Genetic toxicology. I. Rossman,
  Toby G.
  QH465.C5I53   1992
  575.1′31—dc20   92-19547
  ISBN 1-56032-272-1   CIP

Contents

v

Contributors

MIRIT I. ALADJEM
Department of Cell Research &
  Immunology
George S. Wise Faculty of Life Sciences
Tel Aviv University
Ramat Aviv, 69978, Tel Aviv, Israel

HANS PETER AUER
Kerforschungszentrum Karlsruhe
Institut fur Genetik und Toxikologie
D-7500 Karlsruhe 1, Germany

KATHLEEN DIXON
Department of Environmental Health
University of Cincinnati Medical Center
Cincinnati, OH 45267-0056

STEPHEN J. ELLEDGE
Department of Biochemistry
Baylor College of Medicine
Houston, TX 77030

JOSEPH FARGNOLI
Laboratory of Molecular Genetics
National Institute on Aging, National
  Institute of Health
Gerontology Research Center
Baltimore, MD 21224

STEPHAN GEBEL
Kerforschungszentrum Karlsruhe
Institut fur Genetik und Toxikologie
D-7500 Karlsruhe 1, Germany

JUDY HAMMELBURGER
Molecular Carcinogenesis Program
American Health Foundation
Valhalla, NY 10595

PETER HERRLICH
Kerforschungszentrum Karlsruhe
Institut fur Genetik und Toxikologie
D-7500 Karlsruhe 1, Germany

STEVEN HIRSCHFELD
National Institutes of Health, NICHD
Bethesda, MD 20892

NIKKI J. HOLBROOK
Laboratory of Molecular Genetics
National Institute on Aging, National
    Institute of Health
Gerontology Research Center
Baltimore, MD 21224

HARALD KÖNIG
Kerforschungszentrum Karlsruhe
Institut fur Genetik und Toxikologie
D-7500 Karlsruhe 1, Germany

MARCUS KRÄMER
Kerforschungszentrum Karlsruhe
Institut fur Genetik und Toxikologie
D-7500 Karlsruhe 1, Germany

FRANÇOISE LAVAL
Groupe Radiochimie d l'ADN
Unite 247 INSERM
Institut Gustave Roussy
94805 Villejuif, France

SARA LAVI
Department of Cell Research &
    Immunology
George S. Wise Faculty of Life Sciences
Tel Aviv University
Ramat Aviv, 69978, Tel Aviv, Israel

JENNIFER D. LEUTHY
Laboratory of Molecular Genetics
National Institute on Aging, National
    Institute of Health
Gerontology Research Center
Baltimore, MD 21224

ARTHUR S. LEVINE
Laboratory of Developmental and
    Molecular Immunity
National Institute of Child Health and
    Human Development
Bethesda, MD 20892

CHRISTINE LÜCKE-HUHLE
Kernforschungszentrum Karlsruhe
Institut fur Genetik und Toxikologie
D-7500 Karlsruhe 1, Germany

MARY McLENIGAN
Laboratory of Developmental and
    Molecular Immunity
National Institute of Child Health and
    Human Development
Bethesda, MD 20892

MAURO MEZZINA
Laboratory of Molecular Genetics
Institut de Recherches Scientifiques sur
    le Cancer
94801 Villejuif, France

JONG SUNG PARK
Laboratory of Molecular Genetics
National Institute on Aging, National
    Institute on Health
Gerontology Research Center
Baltimore, MD 21224 .

K. ERIC PAULSON
Tufts University
Genetics Department HNRC
Boston, MA 02111

MIROSLAVA PROTIĆ
Laboratory of Developmental and
    Molecular Immunity
National Institute of Child Health and
    Human Development
Bethesda, MD 20892

ADRIANA RADLER-POHL
Kerforschungszentrum Karlsruhe
Institut fur Genetik und Toxikologie
D-7500 Karlsruhe 1, Germany

HANS J. RAHMSDORF
Kerforschungszentrum Karlsruhe
Institut fur Genetik und Toxikologie
D-7500 Karlsruhe 1, Germany

RICHARD ROBINSON
Molecular Carcinogenesis Program
American Health Foundation
Valhalla, NY 10595

ZEEV RONAI
Molecular Carcinogenesis Program
American Health Foundation
Valhalla, NY 10595

SUSAN RUTBERG
Molecular Carcinogenesis Program
American Health Foundation
Valhalla, NY 10595

CHRISTOPH SACHSENMAIER
Kerforschungszentrum Karlsruhe
Institut fur Genetik und Toxikologie
D-7500 Karlsruhe 1, Germany

ALAIN SARASIN
Laboratory of Molecular Genetics
Institut de Recherches Scientifiques sur
   le Cancer
94801 Villejuif, France

BERND STEIN
Kerforschungszentrum Karlsruhe
Institut fur Genetik und Toxikologie
D-7500 Karlsruhe 1, Germany

ALICE P. TSANG
Laboratory of Developmental and
   Molecular Immunity
National Institute of Child Health and
   Human Development
Bethesda, MD 20892

MIRKO VANETTI
Kerforschungszentrum Karlsruhe
Institut fur Genetik und Toxikologie
D-7500 Karlsruhe 1, Germany

ZHENG ZHOU
Department of Biochemistry and
   Institute for Molecular Genetics
Baylor College of Medicine
Houston, TX 77030

Preface

All organisms respond to adverse changes in their environment by eliciting a genetically-programmed chain of events resulting in *de novo* synthesis of a variety of new proteins. The types of proteins induced depend on the nature of the stress, although there are some overlapping responses. Exposure of cells to temperatures higher than normal or to arsenite causes the induction of heat shock proteins (reviewed in Lindquist, 1986), exposure to heavy metal ions leads to the induction of metallothioneins (reviewed in Hamer, 1986), and exposure to oxidative stress leads to the induction of, among other proteins, heme oxygenase (Applegate et al., 1991). These systems are not included in this monograph. DNA damage-inducible genes can be thought of as a subset of stress-responsive genes. The *E. coli* SOS response (see Walker, 1985) has stimulated the search for similar systems in eukaryotes, and for the last ten years there has been an accumulation of new information concerning DNA-damage-inducible processes in eukaryotic cells. The time seems ripe for a monograph on this important topic.

It is now clear that there are some differences between the prokaryotic SOS system and the response of eukaryotic cells to DNA damaging agents. Aside from differences in the mode of control (see Krämer et al., 1990), there is another major difference. The SOS response is required (in *E. coli* and *S. typhimurium*) for mutagenesis to occur in situations where the DNA damage results in a block to replication (by pyrimidine dimers, for example). Thus, cell survival is enhanced at the expense of increased mutation frequency. So far, evidence suggests that this is not the case in eukaryotic cells, which appear to have constitutive systems for mutagenesis (reviewed in Rossman and Klein, 1988).

It is clear that changes in the concentrations or activities of specific protein molecules play central roles in the toxicological response of cells to chemical and radiation damage to DNA. The importance of understanding the significance of results obtained in bioassays for safety assessment makes studies of these inducible changes of crucial importance. Inducible responses to DNA damage may not always be benign. Whereas in some cases (e.g., induction of enzymes involved in DNA repair), the new proteins can be seen as protecting cells from toxic and genotoxic consequences, in other cases (e.g., proteins that lead to gene amplification, and overexpression of oncogenes), the induced protein(s) may represent early biosynthetic perturbations that will eventually lead to carcinogenesis. Identification of inducible proteins may be crucial in helping to explain target organ toxicity and species differences. Understanding the consequences of the induced proteins should be helpful in creating proper mathematical models for extrapolation to low doses. In addition, biomarkers of exposure are gaining importance as tools in molecular epidemiology. Since *de novo* synthesis of certain proteins can be detected after exposure to many xenobiotics, the detection of these proteins or their mRNAs in humans may be useful in assessing human exposure or cellular injury.

Toby G. Rossman, PhD

REFERENCES

Applegate, L. A., Janscher, P., and Tyrell, R. M. (1991). Induction of heme oxygenase: A general response to oxidative stress in cultured mammalian cells. *Cancer Res* 51:974–978.

Hamer, D. H. (1986). Metallothionein. *Ann Rev Biochem* 55:913–951.

Krämer, M., Stein, B., Mai, S., Kunz, E., König, H., Loferer, H., Grunicke, H. H., Ponta, H., Herrlich, P., and Rahmsdorf, H. J. (1990). Radiation-induced activation of transcription factors in mammalian cells. *Radiat Environ Biophys* 29:303–313.

Lindquist, S. (1986). The heat shock response. *Annu Rev Biochem* 55:1151–1191.

Rossman, T. G., and Klein, C. B. (1988). From DNA damage to mutation in mammalian cells. *Environ Mol Mutagen* 11:119–123.

Walker, G. C. (1985). Inducible DNA repair systems. *Annu Rev Biochem* 54:425–457.

1 | XENOBIOTIC REGULATION OF GLUTATHIONE S-TRANSFERASE Ya GENE EXPRESSION

K. Eric Paulson

Genetics Laboratory, USDA Human Nutrition Research
Center on Aging at Tufts University,
Boston, Massachusetts

INTRODUCTION

The glutathione S-transferases (GSTs) represent a family of proteins that conjugate glutathione to a variety of electrophiles. Furthermore, numerous hydrophobic substances are bound with high affinity to GSTs (Pickett and Lu, 1989). The GSTs comprise a family of homo- and hetero-dimers consisting of at least seven subunits. Many of these subunits are expressed in a tissue-specific manner, and several are induced by various xenobiotic compounds that are also substrates for the enzymes (Pickett and Lu, 1989). The induction of GSTs is thought to be an important mechanism of protection against chemical carcinogenesis (Prochaska et al., 1985; Prochaska and Talalay, 1988; Talalay et al., 1988).

CLASSES OF INDUCING AGENTS

The GSTs are part of a group of coregulated enzymes, termed Phase II enzymes, which process electrophiles. Numerous chemical agents have been shown to induce Phase II enzymes, including polycyclic aromatics, phenolic antioxidants, flavenoids, azo dyes, indoles, diterpenes, and isothiocyanates, among others (Huggins, 1979; Wattenberg, 1978, 1983, 1985). Interestingly, many of the agents that induce Phase II enzymes are protective against the toxic and neoplastic effects of carcinogens (Prochaska et al., 1985; Prochaska and Talalay, 1988; Talalay et al., 1988). However, a subset of these chemical agents is also capable of inducing Phase I enzymes (which include several members of the cytochrome P-450 family). Unlike Phase II enzymes, however, Phase I enzymes are thought to be the principal activators of carcinogens to their reactive forms. This subset of chemical agents that induces both the

The author thanks Dr. James E. Darnell, Jr., and Dr. Amy S. Yee for many helpful discussions and for review of this manuscript. This project was funded at least in part with federal funds from the U.S. Department of Agriculture, Agricultural Research Service, under contract 53-3K06-0-1. The contents of this publication do not necessarily reflect the views or policies of the U.S. Department of Agriculture.

Phase I and II enzymes consists of large planar aromatics such as the polycyclic aromatics, flavenoids, azo dyes, and tetrachlorodibenzodioxin (TCDD). Thus, the anticarcinogenic chemicals can be classified into bifunctional inducers, which elevate Phase I and II enzymes, and monofunctional inducers, which elevate Phase II enzymes without inducing Phase I enzymes.

Talalay has proposed a model for the regulation of Phase II enzymes (including GSTs) by both monofunctional and bifunctional inducers (Prochaska et al., 1985; Prochaska and Talalay, 1988; Talalay et al., 1988). Bifunctional inducers activate Phase II enzymes via two pathways. The first pathway involves the aryl hydrocarbon (Ah) receptor, which binds to specific DNA sequences and induces transcription of both Phase I and Phase II genes. In addition, bifunctional inducers can be metabolized first by the induced cytochrome P-450s and are then converted to monofunctional inducing agents. The monofunctional inducers, which activate Phase II enzymes only, utilize a signal that is independent of the Ah receptor or metabolism by P-450s. As described in the following sections, the molecular analysis of the GST Ya gene provides evidence for this model.

THE MECHANISMS OF INDUCTION OF THE GST Ya GENE BY XENOBIOTICS

The mechanisms underlying the regulation of the GSTs are important to understanding the metabolism of xenobiotics. Pickett's group and others have demonstrated that the xenobiotic-induced increase in rat liver GST Ya protein levels following induction is correlated with an equal induction in rat liver GST Ya mRNA (Lai et al., 1984; Pickett et al., 1984). Importantly, this showed that GST Ya levels are regulated pretranslationally, which suggested that the GST Ya gene might be transcriptionally regulated. Analysis of the transcription rate of the rat GST Ya gene in liver using nuclear runon assays showed approximately a four- to fivefold increase upon induction with 3-methylcholanthrene (3-MC) (Fig. 1) (Ding and Pickett, 1985). This indicates that much of the regulation by xenobiotics occurs at the transcriptional level, but that there is also likely a component of posttranscriptional regulation such as message stabilization. A similar combination of transcriptional and posttranscriptional regulation has been observed in the regulation of mouse liver GST Ya gene (Pearson et al., 1988). It should also be noted that some xenobiotic-induced cytochrome P-450 genes are also regulated by both transcriptional and posttranscriptional components (Kimura et al., 1986; Pasco et al., 1988). This suggests that this combination of regulatory mechanisms may be a general feature of xenobiotic inducibility. Although the role of xenobiotic compounds in the stabilization of GST Ya mRNA is not clear,

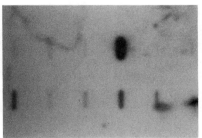

**5 DAYS
SALINE**

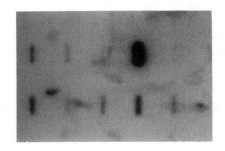

**5 DAYS
3-MC**

PATTERN

GST   LIV3   LIV5   ALB   AFP   PAH

GS   βTUB   ACT   tMET   CHOB pBR

FIGURE 1. Effect of 3-MC on transcription rates of the GST Ya gene in rat liver. In vitro runon experiments were performed on rats that had been treated with 3-MC for 5 d or on control rats that received saline for 5 d. Labeled RNA (4 × 10^7 cpm per sample) was hybridized under conditions of DNA excess to cDNA inserts of the following genes: GST (glutathione S-transferase Ya), GS (glutamine synthetase), LIV3 (mouse α-1 antitrypsin), LIV 5 (mouse transthyretin), ALB (mouse albumin), AFP (mouse alpha-fetoprotein), PAH (phenylalanine hydroxylase), β-TUB (mouse β-tubulin), ACT (chicken actin), tMET (methionine tRNA), CHOB (Chinese hamster ovary clone B, a control), pBR (pBR 322, a control). Filters were washed, treated with RNAse A, and subjected to autoradiography.

as shown in the next section, xenobiotics have a direct role in regulating transcription.

Transcriptional Regulation by Bifunctional Inducers

To understand xenobiotic transcriptional regulation, the GST Ya promoter must be defined with respect to cis-acting elements and the transcriptional factors that interact with these elements. Such an analysis

provides a framework for understanding how GST Ya gene transcription
is modulated by inducers in cooperation with basal transcription. In a
standard approach to identifying DNA sequences involved in both inducible and constitutive expression, deletion mutants in the upstream
regions of the GST Ya gene were constructed. These deletion mutants
were then transfected into the human hepatoma cell line HepG2 and
tested for both basal activity and induction by the bifunctional agent 3-
MC.

The deletion analysis of the GST Ya gene showed that all of the elements required for both basal and inducible activity lie between −980
and −650 bp (Fig. 2) (Paulson et al., 1990). Deletions that retained up to
980 bp of upstream sequence also retained full basal activity and 3-MC
inducible activity, while deletions to −650 bp or less had no inducible or
basal activity. Deletion to −875 bp had little effect on basal activity but

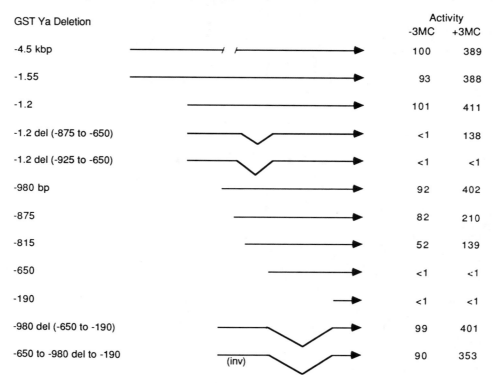

FIGURE 2. GST Ya enhancer defined by deletions of the GST Ya upstream region. The size of
progressive upstream deletions is indicated on the left and is graphically illustrated in the center.
Internal deletions are indicated by del ( ) and are also graphically illustrated by the bent bridge.
The transcriptional activity of the constructs ± 3-MC induction is represented at the right. Each
experiment was normalized to a β-globin transfection control signal before comparison for GST Ya
expression. The full upstream sequence of −4.5 kbp is taken as 100% activity without induction.
Each point is the average of at least two individual experiments.

lost about half of the induced activity, indicating a possible inducible element. Further internal deletions of the plasmid with −1.2 kbp upstream located the inducible element. A deletion between −875 and −650 bp retained some inducibility, although no basal activity was now detectable. Expanding this deletion by only 50 additional bases (total deletion −925 to −650 bp) completely eliminated 3-MC inducibility. These deletions definitively demonstrated that a 3-MC-inducible element lies between −925 and −875 bp and also indicated that all basal elements resided between −875 and −650 bp. Additional deletions showed that sequences between −875 and −650 bp contained both basal and additional inducible activity. Further deletions by Pickett's group (Rushmore et al., 1990) localized a second inducible element between −722 and −689 bp.

Taken together, the 5′ deletion data indicated that the sequences from −980 to −650 bp comprised an upstream enhancer and were required for full activity of the GST Ya gene. As a further test of the enhancer activity of the −980 to −650 bp region, we required that it must also function on a heterologous promoter. The GST Ya enhancer was in fact shown to be capable of activating the β-globin promoter three- to fivefold in untreated HepG2 cells and 10- to 20-fold in the presence of 3-MC (Paulson et al., 1990). This is approximately the same degree of stimulation by 3-MC that was observed in the natural gene and confirmed that the −980 to −650 bp region was necessary and sufficient for both basal and inducible enhancer activity.

The nature of the 3-MC-inducible sequence between −925 and −875 bp was demonstrated to likely be involved in interaction with the Ah or dioxin receptor (Paulson et al., 1990). In order to identify the specific sites within the −925 to −875 bp region that were involved in 3-MC induction, nuclear extracts from HepG2 cells that were treated or untreated with 3-MC were used to examine protein-DNA contacts. Protein-DNA binding was tested by electrophoretic mobility shift assays where protein-DNA complexes are resolved from unbound labeled DNA. It was found that one or more proteins specifically bind to the DNA probe (−980 to −875 bp) in both 3-MC-treated and untreated extracts (Fig. 3a). The specific complexes are identified by the ability of excess homologous cold DNA to compete with the gel shift bands. Numerous oligonucleotides encompassing other protein binding sites were unable to compete with these bands. Two bands are present in all extracts (identified as constitutive or C), while two other bands are present only in 3-MC-treated extracts (identified as induced or I). The induced bands appear within 30 min after 3-MC treatment but are diminished after 4 h. Furthermore, extracts made after 8 or 24 h of 3-MC treatment showed only the constitutive bands. The rapid induction of the protein responsible for the gel shift bands correlated well with the rapid induction of the transfected gene (induction could be observed within 2 h, the shortest time

GST Ya Probe: -980 to -875

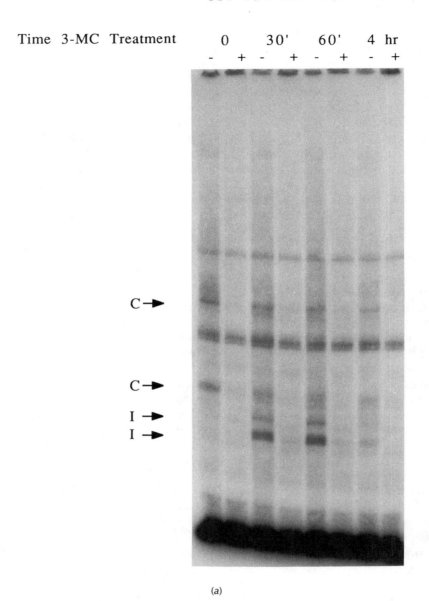

(a)

FIGURE 3. Analysis of the 3-MC-inducible activation region −925 to −875 bp. (a) Nuclear protein-DNA interactions in the 3-MC-inducible region. The GST Ya upstream segment −980 to −875 bp was 5′ end-labeled and incubated with untreated or 3-MC-treated HepG2 nuclear extracts in the presence (+) or absence (−) of a 40-fold molar excess of cold homologous DNA. The binding reaction was analyzed by a gel retardation assay in which protein-DNA complexes are separated from free DNA on a low-ionic-strength 4% polyacrylamide gel. Specifically competed bands that do not vary with 3-MC treatment are labeled C for constitutive. Specifically competed bands that do vary with 3-MC treatment are labeled I for induced.

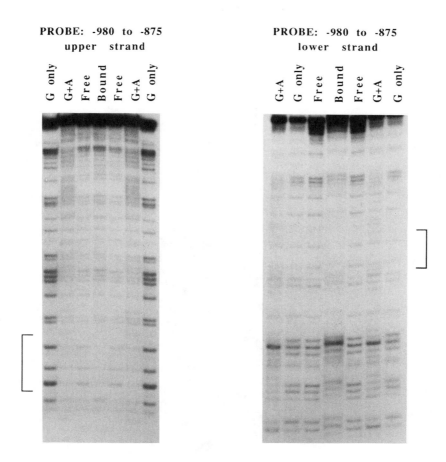

PROBE: -980 to -875
upper strand

PROBE: -980 to -875
lower strand

```
                    +   + +
  -9 0 5   CGTCAGGCATGTTGCGTGCA - 8 8 5
           GCAGTCCGTACAACGCACGT
                +     +
```

(b)

FIGURE 3. Analysis of the 3-MC-inducible activation region −925 to −875 bp (*Continued*). (*b*)Analysis of the nucleotide residues involved in 3-MC-induced protein binding by DMS-methylation protection. The standard binding reaction was scaled up 20-fold and was treated with dimethyl sulfate before being loaded on a preparative 4% polyacrylamide gel. Both 5′ and 3′ end-labeled fragments (−980 to −875 bp) were used in binding reactions in order to identify protected residues on both strands. Bound DNA bands were isolated and cleaved with piperidine and analyzed on a sequencing gel. The brackets indicate the G residues from the protein-bound bands that were protected from methylation. The bottom of the figure indicates the positions of the protected G residues.

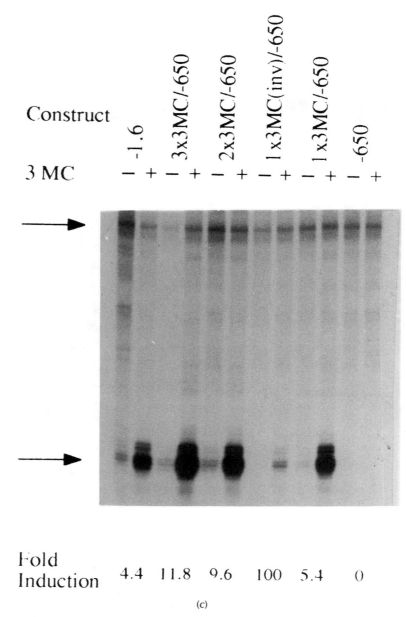

FIGURE 3. Analysis of the 3-MC-inducible activation region −925 to −875 bp (*Continued*). (*c*) Functional assay of the 3-MC-inducible protein-DNA interaction. A double-stranded oligonucleotide encompassing the methylation protected sequence (−910 to −875 bp) was prepared and cloned in various multimers as indicated in front of the GST Ya promoter at −650 bp. The construct was transfected into HepG2 cells and the transcriptional activity ± 3-MC was assayed. The upper arrow indicates the β-globin first exon (transfection control) and the lower arrow indicates the GST Ya first exon.

examined), suggesting the involvement of this protein in induction of the GST Ya gene.

To locate precisely the DNA binding site for the induced protein factor, DMS-methylation protection footprints were performed on the induced protein-DNA complexes (Paulson et al., 1990). Protected G residues were found on both the upper and lower strand between −905 and −885 bp, that is, within the region identified as necessary in the functional analysis (−925 to −875 bp) (Fig. 3b). Finally, to test the functionality of this protected sequence a double-stranded oligonucleotide that encompassed −910 to −875 bp was prepared and cloned in front of the GST Ya promoter at −650 bp. As shown in Fig. 3c, the promoter plus the −910 to −875 bp oligonucleotide was induced by 3-MC, while the construct without the upstream element was not. Figure 3c also demonstrates that the effect of multiple binding sites is nearly additive. Furthermore, as expected, the −910 to −875 bp oligonucleotide was able to specifically complete the induced gel shift bands formed with the −980 to −875 bp probe.

Although the results of these experiments identified a 3-MC-inducible positive-acting element, to which a 3-MC-inducible protein factor binds, the nature of the protein factor was not directly demonstrated. However, we noted that the sequence of the footprinted region of the GST Ya gene resembles the consensus sequence of the xenobiotic response elements or XREs (shown below) first described in the inducible cytochrome P-450 genes (Fisher et al., 1990; Frilling et al., 1990; Fujisawa-Sehara et al., 1986; Gonzalez and Nebert, 1985; Jones et al., 1985; Neuhold et al., 1989; Sasawa et al., 1986). Furthermore, the XREs are inducibly bound by the Ah or dioxin receptor (Denison et al., 1988; Fujisawa-Sehara et al., 1988; Hapgood et al., 1989; Saatcioglu et al., 1990; Wilhelmsson et al., 1990).

GST Ya:
```
          ⌒➤
CAGGCATGTTGCGTGCAT
GTCCGTACAACGCACGTA
    ◄⌒
```

Consensus:
```
     GTTGCGTGA
     C A     C
```

The binding site of the 3-MC-inducible protein in the GST Ya gene has several similarities to the various steroid receptor binding sites, including an inverted repeat in the contact region (with one base pair mismatch). This characteristic was not previously described for XREs. However, because the Ah receptor has been postulated to be a member of the steroid receptor superfamily (Hapgood et al., 1989; Nemoto et al., 1990; Saatcioglu et al., 1990; Wilhelmsson et al., 1990), it might be predicted that XREs could have a structure similar to steroid receptor binding sites (Evans, 1988). Mutagenesis studies are currently being pursued to determine which bases are most important for induced protein binding and xenobiotic inducible function and to determine whether the observed dyad symmetry in the binding site has a functional role.

Additional evidence suggests that the Ah receptor may be the 3-MC-inducible protein detected in our assays with the GST Ya XRE. The kinetics of 3-MC induction of the GST Ya protein-DNA complex are very similar to the induction kinetics of the Ah receptor. Only 1 h of exposure to TCDD is required for maximal induction of the Ah receptor that binds a P-450 XRE. In addition, this binding activity is greatly diminished at 4.5 h of exposure and is completely gone by 16 h (Hapgood et al., 1989). It should also be noted that Pickett's group has shown that the GST Ya gene is not responsive to xenobiotics when transfected into cell lines deficient in the Ah receptor (Telakowski-Hopkins et al., 1988). Finally, we have shown that cycloheximide has no effect on 3-MC induction of the GST Ya gene, indicating that the 3-MC-induced protein-DNA interaction may be due to the preexisting Ah receptor (Paulson et al., 1990). More direct correlations await the availability of an antibody reagent directed against the Ah receptor.

As shown in Fig. 3a, both constitutive and inducible proteins interact with the 3-MC inducible protein, because an oligonucleotide encompassing the sequence competes with the constitutive gel shift bands (Paulson et al., 1990). We have not identified the nature of the constitutive binding proteins. However, the possibility of multiple proteins interacting at one binding site is not unprecedented, especially within the steroid receptor superfamily. The retinoic acid and thyroid hormone receptors appear to activate through a common sequence element (Umesono et al., 1988). Also, the thyroid hormone receptor appears to act as an antagonist to estrogen receptor binding at the same site and consequently blocks transcriptional induction by estrogen (Glass et al., 1988). Furthermore, the AP-1 transcription factor appears to be able to competitively bind to glucocorticoid receptor binding sites (Diamond et al., 1990). We don't have any evidence that the constitutive binding proteins function mechanistically as in these examples. However, if the xenobiotic receptor is a member of the steroid receptor family then there are several possible mechanisms to test for the function of the constitutive proteins.

Transcriptional Regulation by Monofunctional Inducers

Although the upstream inducible element has the hallmarks of an XRE, the sequence between -722 and -689 bp defined by Pickett does not have any sequence homology to XREs (Paulson et al., 1990; Rushmore et al., 1990). However, it does contribute to maximum inducible expression (Fig. 1) (Paulson et al., 1990; Rushmore et al., 1990). In a recent study,

Pickett's group demonstrated that the −722 to −689 bp element can be functionally distinguished from the XRE by its response to phenolic antioxidants (Rushmore and Pickett, 1990). While both elements respond to bifunctional compounds such as 3-MC and β-naphthoflavone (βNF), only the −722 to −689 bp element responds to tert-butylhydroquinone (tert-BHQ), which is a phenolic antioxidant in the monofunctional inducer class. This element has therefore been termed the antioxidant response element or ARE. In addition, the ARE does not require a functional Ah receptor or the cytochrome P-450 IA1 protein when induced by tert-BHQ. These experiments confirm the model proposed by Talalay, in which an Ah receptor-independent mechanism can account for the induction of GSTs and other Phase II enzymes by monofunctional inducers (Prochaska et al., 1985; Prochaska and Talalay, 1988; Talalay et al., 1988). If the ARE in the GST Ya gene represents a general mechanism, it may be present in the regulatory regions of other Phase II genes as a means of coordinate regulation by monofunctional inducers, similar to the function of XRE sequences.

The nature of the protein factor that interacts with the ARE is currently unknown, although a protein factor binds within the region (Paulson et al., 1990; Rushmore et al., 1990). When a DNA probe from −725 to −650 bp is used in a gel shift assay, a specific and compatible band was apparent (Fig. 4a). In addition, the location of the protein-DNA binding site was determined by dimethyl sulfate (DMS)-methylation protection footprinting (Fig. 4b). The footprints of the upper and lower strand show that the site of the interaction is between −700 and −680 bp, which is within the region of the ARE. However, there is no variability in the binding of this band with or without treatment by xenobiotics (Paulson et al., 1990; Rushmore et al., 1990). Furthermore, this complex was competed by a fragment from the albumin enhancer that is bound by a positive acting protein, termed NLS (Herbst et al., 1989). The NLS factor that binds a positive-acting site in the albumin enhancer is not cell specific in its distribution, and similarly the distribution of the protein that formed complexes on the −725 to −650 bp probe was also present in several cell types (Paulson et al., 1990). The possible identity of the NLS factor and the GST Ya binding protein was examined by cross-competition (Fig. 4a). Both the unlabeled NLS fragment from the albumin enhancer and the unlabeled −725 to −650 bp fragment were able to compete with a gel shift band produced by a partly purified NLS factor. Comparison of this binding site and the albumin enhancer binding site does not show any obvious homologies. However, it may require several additional footprints of other sites before a consensus sequence can be determined. While the footprinted sequences are within the region defined as an ARE, it is unclear whether the NLS-like protein that interacts with the sequence is involved in regulation by monofunctional inducers. Clearly the protein is not cell-type specific and its binding is not regulated by xenobiotics. However, that does not mean that its transcriptional activity is not regulated. A functional test of the transcriptional properties of the NLS-like protein might be possible using an in vitro cell-free system, but that type of experiment typically requires purified protein, which is currently unavailable.

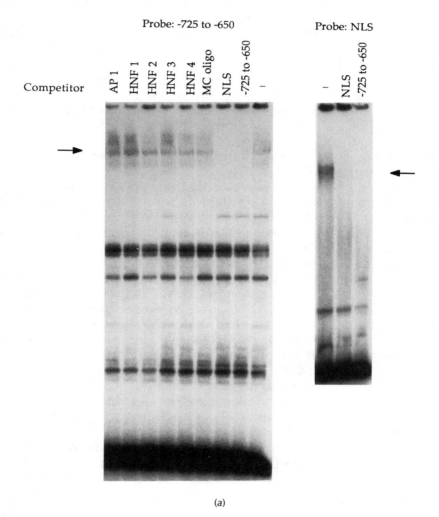

(a)

FIGURE 4. Nuclear protein-DNA interactions in the region −725 to −650 bp. (a) The left side of the figure shows DNA fragment −725 to −650 bp, which was 5′ end-labeled and incubated with HepG2 nuclear extracts. The right side of the figure shows the NLS fragment from the albumin enhancer, which was 5′ end-labeled and incubated with partially purified NLS protein. Reactions included either no cold competitor DNA or a 40-fold molar excess of cold competitor DNA as indicated. The binding reaction was assayed on a nondenaturing gel as in Fig. 5. The specifically competed NLS shift is indicated with an arrow.

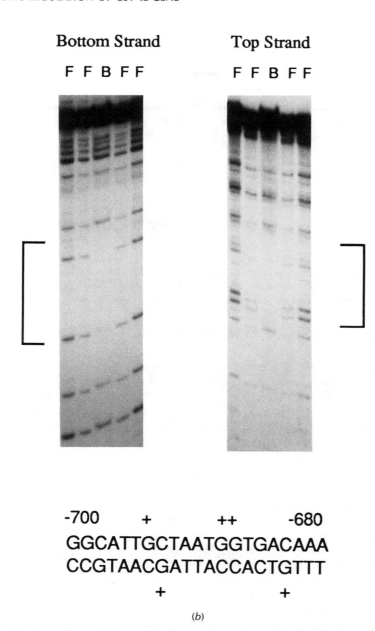

Bottom Strand   **Top Strand**

F F B F F   F F B F F

-700   +   ++   -680
GGCATTGCTAATGGTGACAAA
CCGTAACGATTACCACTGTTT
        +   +

(b)

FIGURE 4. Nuclear protein-DNA interactions in the region −725 to −650 bp (*Continued*). (*b*) Analysis of the nucleotide residues involved in the NLS interaction. DMS-methylation protection was performed as described in Fig. 3. The brackets indicate the G residues from the protein-bound bands that were protected from methylation. The bottom of the figure indicates the location of the protected G residues.

-950   -650

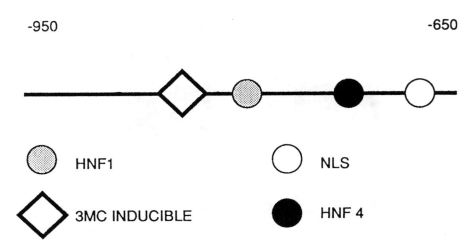

FIGURE 5. Summary of the nuclear protein binding sites regulating the function of the GST Ya enhancer. Shown schematically is the GST Ya enhancer from −950 to −650 bp. The different circles or boxes represent the various protein-DNA interactions as indicated.

THE MECHANISMS OF BASAL EXPRESSION

Analysis of the basal elements and factors of the GST Ya gene has also provided clues into the tissue-specific nature of this promoter. As described earlier (Fig. 1) (Paulson et al., 1990), constructs containing the sequences between −875 and −650 bp are required for normal basal gene expression. In order to identify specific sites involved in basal expression, gel-shift experiments were performed using DNA probes covering −875 to −726 bp. As shown in Fig. 5, which summarizes the protein-DNA interaction data, the constitutive hepatocyte enriched factors HNF1 (Courtois et al., 1988) and HNF4 (Costa et al., 1989) were identified by gel-shift binding analysis and confirmed by DMS-methylation protection footprinting and Exo III assays of the protein-DNA complexes (Paulson et al., 1990). HNF1 is known to be a positive-acting factor, and the deletion of this site decreased HepG2 expression of the GST construct about twofold (Fig. 1). In addition, deletion of HNF1 sites from several other liver-specific enhancers has resulted in similar or even smaller but reproducible reductions in enhancer activity in HepG2 cells (Costa et al., 1989; Grayson et al., 1988). Similarly, the HNF4 protein has been shown to be a positive-acting factor in several liver-specific genes (Costa et al., 1989; Grayson et al., 1988). Interestingly, the arrangement of liver-specific and constitutive factor binding sites seems to be a general feature of several liver-specific enhancers and promoters studied so far (Babiss et al., 1987; Cereghini et al., 1987; Costa et al., 1987; Grayson et al., 1988; Lichsteiner et al., 1987; Maire et al., 1989). The GST Ya enhancer also fits this pattern in its arrangement of binding proteins.

INDUCIBLE TISSUE-SPECIFIC AND POSITION-SPECIFIC REGULATION OF THE GST Ya GENE IN VIVO

Having identified the critical sequences and protein-binding sites for both basal and 3-MC-inducible expression of the GST Ya gene in vitro, we have begun assessing the role of these sequences in vivo. We have produced transgenic mice to examine the role of upstream GST Ya sequences on inducible expression in vivo and the role of the constitutive elements in general expression. Several lines of transgenic mice carrying −1.55 kbp of upstream sequence driving the herpes TK gene are fully capable of being induced by 3-MC (K. E. Paulson and J. E. Darnell, manuscript in preparation). Interestingly, these transgenic animals are incapable of being induced by phenobarbital, although the endogenous mRNA is induced. This could mean that we have not included sufficient DNA sequences in our constructs, or that phenobarbital does not regulate GST Ya transcriptionally.

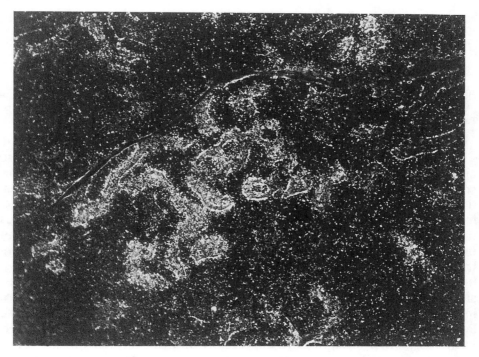

FIGURE 6. In situ hybridization of 3-MC-induced mouse kidney tissue. Female CD1 induced 48 h with 3-MC, dark-field exposure using a GST Ya probe. Magnification, ×100. Kidney tissue was prepared for in situ hybridization by fixing overnight in 4% paraformaldehyde in 1 × PBS at 4°C. The tissue was then equilibrated overnight in 30% sucrose at 4°C and mounted in Tissue-Tek O.C.T. compound for frozen sectioning; 5-μm sections were hybridized with an antisense ^{35}S-labeled probe prepared from a T7 RNA transcript of a rat GST Ya cDNA. Sections were washed and treated with RNAse A, then exposed on Kodak NTB-2 emulsion for 2–4 d.

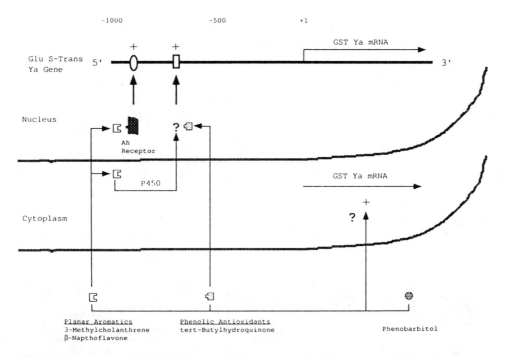

FIGURE 7. Summary of the mechanisms of induction by monofunctional and bifunctional xenobiotic inducers. Shown schematically are the GST Ya gene and the sites at which regulation occurs. The oval to the left represents the XRE while the rectangle to the right represents the ARE.

The tissue distribution and xenobiotic regulation of GST Ya in mouse and rat are predominantly in liver and to a lesser extent in kidney (Pearson et al., 1988). Interestingly, it has been demonstrated that the regulation of the GST Ya gene is even more complicated than regulating inducible tissue-specific expression. Using in situ hybridization, it was demonstrated that the GST Ya mRNA is present in all hepatocytes in uninduced mice but is specifically induced by xenobiotics in the pericentral region of the liver (Paulson et al., 1990). Pericentral expression following induction is sharply demonstrated. In addition, pericentral expression is characteristic for several other enzymes involved in detoxification, including some cytochrome P-450 isozymes and epoxide hydrolase (Moody et al., 1983; Ratanasavanh et al., 1986; Wolf et al., 1984). Furthermore, as shown in Fig. 6, the expression of GST Ya mRNA in kidney is highly localized as well. Expression appears to be specific for proximal tubules in the kidney cortex following induction with 3-MC. Interestingly, the expression of GST Ya in both kidney and liver is colocalized with mouse major urinary protein (MUP) and glutamine synthetase RNA (Kuo et al., 1988; Paulson et al., 1990; K. E. Paulson, unpublished data). Such all-or-none positional expression for two very different genes seems unlikely to be due to gradients of metabolites as previously postulated (Jungermann, 1986). Therefore, region-specific distribution patterns most likely represent positional regulation of gene expression within the same cell lineage due to other signals, for example, extracellular cell surface contacts. Based on the in situ hybridization results on xenobiotic induction and the identification of two different regulatory elements, which control induction, we have speculated that these sites and their transacting factors might mediate region-specific regulation. We are unable to test this hypothesis directly because antibody probes to the transacting factors (e.g., the Ah receptor) are not available. However, experiments are in progress to examine the function of the inducible elements in pericentral localization in the liver.

SUMMARY AND FUTURE DIRECTIONS

The pioneering studies of Talalay on the regulation of Phase II proteins by xenobiotic compounds and the model that came from those studies have in large part been confirmed by the molecular analysis of the GST Ya gene. As shown in Fig. 7, this model consists of two independent response elements in the GST Ya gene. The upstream element is an XRE and responds only to bifunctional inducers via the Ah receptor. The second element is an ARE that responds to both monofunctional and bifunctional inducers via an Ah-receptor-independent mechanism.

While a consensus sequence for XREs has been identified, it is unknown which bases are required for function and interaction with the Ah receptor. Mutagenesis studies are currently underway that are examining the GST Ya XRE in detail, including the role of dyad symmetry in the binding site. Additional mutagenic studies are also required to more precisely identify the specific nucleotides in the ARE that are required for induction. Furthermore, are these sequences part of the binding site of the NLS-like protein, and if so what is the mechanism for activation?

Finally, the in vivo regulation of the GST Ya gene by xenobiotics may provide a clue to the development of the mammalian liver. The induction by xenobiotics could allow us to determine the transcriptional elements and their transacting factors that are involved in positional regulation. This information might enable us in turn to learn more about the architecture of the liver and the influence of lineage and extracellular matrix on liver gene expression.

REFERENCES

Babiss, L. E., Herbst, R. S., Bennett, A. L., and Darnell, J. E., Jr. 1987. Factors that interact with the rat albumin promoter are present in hepatocytes and other cell types. *Genes Dev.* 1:256–267.

Cereghini, S., Raymondjean, M., Carranca, A. G., Herbomel, P., and Yaniv, M. 1987. Factors involved in control of tissue-specific expression of albumin gene. *Cell* 50:627–638.

Costa, R. H., Grayson, D. R., and Darnell, J. E., Jr. 1989. Multiple hepatocyte-enriched nuclear factors function in the regulation of transthyretin and α1-antitrypsin genes. *Mol. Cell. Biol.* 9:1415–1425.

Courtois, G., Baumhueter, S., and Crabtree, G. R. 1988. Purified hepatocyte nuclear factor 1 interacts with a family of hepatocyte-specific promoters. *Proc. Natl. Acad. Sci. USA* 85:7937–7941.

Denison, M. S., Fisher, J. M., and Whitlock, J. P., Jr. 1988. Inducible, receptor-dependent protein-DNA interactions at a dioxin-response transcriptional enhancer. *Proc. Natl. Acad. Sci. USA* 85:2528–2532.

Diamond, M. I., Miner, J. N., Yoshinaga, S. K., and Yamamoto, K. R. 1990. Transcription factor interactions: Selectors of positive or negative regulation from a single DNA element. *Science* 249:1266–1272.

Ding, V. D.-H., and Pickett, C. B. 1985. Transcriptional regulation of rat glutathione S-transerase genes. *Arch. Biochem. Biophys.* 240:553–559.

Evans, R. M. 1988. The steroid and thyroid hormone receptor superfamily. *Science* 240:889–895.

Fisher, J. M., Wu, L., Denison, M. S., and Whitlock, J. P., Jr. 1990. Organization and function of a dioxin-responsive enhancer. *J. Biol. Chem.* 265:9676–9681.

Frilling, R. S., Bensimon, A., Tichauer, Y., and Daniel, V. 1990. Xenobiotic-inducible expression of murine glutathione S-transferase Ya subunit gene is controlled by an electrophile-responsive element. *Proc. Natl. Acad. Sci. USA* 87:6258–6262.

Fujisawa-Sehara, A., Sogawa, K., Nishi, C., and Fujii-Kuriyama, Y. 1986. Regulatory DNA elements localized remote upstream from the drug-metabolizing cytochrome P450c gene. *Nucleic Acids Res.* 14:1465–1477.

Fujisawa-Sehara, A., Yamane, M., and Fujii-Kuriyama, Y. 1988. A DNA-binding factor specific for xenobiotic responsive elements of P-450C gene exists as a cryptic form in cytoplasm: Its possible translocation to nucleus. *Proc. Natl. Acad. Sci. USA* 85:5659–5863.

Glass, C. K., Holloway, J. M., Devary, O. V., and Rosenfeld, M. G. 1988. The thyroid hormone receptor binds with opposite transcriptional effects to a common sequence motif in thyroid hormone and estrogen response elements. *Cell* 54:313–323.

Gonzalez, F. J., and Nebert, D. W. 1985. Autoregulation plus upstream positive and negative control regions associated with transcriptional activation of the mouse P1450 gene. *Nucleic Acids Res.* 113:7269–7288.

Grayson, D. R., Costa, R. H., Xanthopoulos, K. G., and Darnell, J. E., Jr. 1988. A cell-specific enhancer of the mouse a1-antitrypsin gene has multiple functional regions and corresponding protein-binding sites. *Mol. Cell. Biol.* 8:1055–1066.

Hapgood, J., Cuthill, S., Denis, M., Poellinger, L., and Gustafsson, J.-A. 1989. Specific-protein DNA interactions at a xenobiotic-responsive element: Copurification of dioxin receptor and DNA. *Proc. Natl. Acad. Sci. USA* 86:60–64.

Herbst, R. S., Friedman, N., Darnell, J. E., and Babiss, L. E. 1989. Positive and negative regulatory elements in the mouse albumin enhancer. *Proc. Natl. Acad. Sci. USA* 86:1533–1557.

Huggins, C. B. 1979. *Experimental Leukemia and Mammary Cancer. Induction, Prevention, Cure.* Chicago: University of Chicago Press.

Jones, P. B. C., Galeazzi, D. R., Fisher, J. M., and Whitlock, J. P., Jr. 1985. Control of cytochrome P1-450 gene expression by dioxin. *Science* 227:1499–1502.

Jungermann, K. 1986. Zonal signal heterogeneity. *Regulation of Hepatic Metabolism*, eds. R. G. Thuman, F. C. Kauffman, and K. Jungermann, pp. 445–472. New York: Plenum.

Kimura, S., Gonzalez, F. J., and Nebert, D. W. 1986. Tissue-specific expression of the mouse dioxin inducible $P_1$450 and $P_3$450 genes: Differential activation and mRNA stability in liver and extra-hepatic tissues. *Mol. Cell. Biol.* 6:1471–1477.

Kuo, C. F., Paulson, K. E., and Darnell, J. E., Jr. 1988. Positional and developmental regulation of glutamine synthetase expression in mouse liver. *Mol. Cell. Biol.* 8:4966–4971.

Lai, H.-C. J., Li, N. L., Weiss, M. J., Reddy, C. C., and Tu, C.-P. D. 1984. The nucleotide sequence of a rat liver glutathione S-transferase subunit cDNA clone. *J. Biol. Chem.* 259:5536–5542.

Lichtsteiner, S., Wuarin, J., and Schibler, U. 1987. The interplay of DNA-binding proteins on the promoter of the mouse albumin gene. *Cell* 51:963–973.

Maire, P., Wuarin, J., and Schibler, U. 1989. The role of cis-acting promoter elements in tissue-specific albumin gene expression. *Science* 244:343–346.

Moody, D. E., Taylor, L. A., Smuckler, E. A., Levin, W., and Thomas, P. E. 1983. Immunohistochemical localization of cytochrome P450a in liver sections from untreated rats and rats treated with phenobarbitol or 3-methylcholanthrene. *Drug Metab. Dispos.* 4:339–343.

Nemoto, T., Mason, G. F., Wilhelmsson, A., Cuthill, S., Hapgood, J., Gustafsson, J.-A., and Poellinger, L. 1990. Activation of the dioxin and glucocorticoid receptors to a DNA binding state under cell-free conditions. *J. Biol. Chem.* 265:2269–2277.

Neuhold, L. A., Shirayoshi, Y., Ozato, K., Jones, J. E., and Nebert, D. W. 1989. Regulation of mouse CYP2A2 gene expression by dioxin: Requirement of two cis-acting elements during induction. *Mol. Cell. Biol.* 9:2378–2386.

Pasco, D. S., Boyum, K. W., Merchant, S. N., Chalberg, S. C., and Fagan, J. B. 1988. Transcriptional and post-transcriptional regulation of the genes encoding cytochromes P-450c and P-450d in vivo and in primary hepatocyte cultures. *J. Biol. Chem.* 263:8671–8676.

Paulson, K. E., Darnell, J. E., Jr., Rushmore, T. H., and Pickett, C. B. 1990. Analysis of the upstream elements of the xenobiotic compound-inducible and positionally regulated glutathione S-transferase Ya gene. *Mol. Cell. Biol.* 10:1841–1852.

Pearson, W. R., Reinhart, J., Sisk, S. C., Anderson, K. S., and Adler, P. N. 1988. Tissue-specific induction of murine glutathione transferase mRNAs by butylated hydroxyanisole. *J. Biol. Chem.* 263:13324–13332.

Pickett, C. B., and Lu, A. Y. H. 1989. Glutathione S-transferases: Gene structure, regulation and biological function. *Ann. Rev. Biochem.* 58:734–764.

Pickett, C. B., Telakowski-Hopkins, C. A., Ding, G. J.-F., Argenbright, L., and Lu, A. Y. H. 1984. Rat liver glutathione S-transferases: Complete nucleotide sequence of a glutathione S-transferase mRNA and the regulation of the Ya, Yb and Yc mRNAs by 3-methylcholanthene and phenobarbitol. *J. Biol. Chem.* 259:5182–5188.

Prochaska, H. J., De Jong, M. J., and Talalay, P. 1985. On the mechanisms of induction of cancer-protective enzymes: A unifying proposal. *Proc. Natl. Acad. Sci. USA* 82:8232–8236.

Prochaska, H. J., and Talalay, P. 1988. Regulatory mechanisms of monofunctional and bifunctional anticarcinogenic enzyme inducers in murine liver. *Cancer Res.* 48:4776–4782.

Ratanasavanh, D., Beune, P., Baffet, G., Rissel, M., Kremers, P., Guengerich, F., and Guillouzo, A. 1986. Immunohistochemical evidence for the maintenance of cytochrome P450 isoenzymes, NADPH cytochrome c reductase and epoxide hydrolase in pure and mixed primary cultures of adult hepatocytes. *J. Histochem. Cytochem.* 34:527–533.

Rushmore, T. H., King, R. G., Paulson, K. E., and Pickett, C. B. 1990. Regulation of glutathione S-transferase Ya subunit gene expression: Identification of a unique xenobiotic-responsive element controlling inducible expression by planar aromatic compounds. *Proc. Natl. Acad. Sci. USA* 87:3826–3830.

Rushmore, T. H., and Pickett, C. B. 1990. Transcriptional regulation of the rat glutathione S-transferase Ya subunit gene. Characterization of a xenobiotic-responsive element controlling inducible expression by phenolic antioxidants. *J. Biol. Chem.* 265:14648–14653.

Saatcioglu, F., Perry, D. J., Pasco, D. S., and Fagan, J. B. 1990. Aryl hydrocarbon (Ah) receptor DNA-binding activity. Sequence specificity and Zn^{+2} requirement. *J. Biol. Chem.* 265:9251–9258.

Sogawa, K., Fujisawa-Sehara, A., Yamane, M., and Fujii-Kuriyama, Y. 1986. Location of regulatory elements responsible for drug induction in rat cytochrome P450c gene. *Proc. Natl. Acad. Sci. USA* 83:8044–8048.

Talalay, P., De Jong, M. J., and Prochaska, H. J. 1988. Identification of a common chemical signal regulating the induction of enzymes that protect against chemical carcinogenesis. *Proc. Natl. Acad. Sci. USA* 85:8261–8265.

Telakowski-Hopkins, C. A., King, R. G., and Pickett, C. B. 1988. Glutathione S-transferase Ya subunit gene: Identification of regulatory elements required for basal level and inducible expression. *Proc. Natl. Acad. Sci. USA* 85:1000–1004.

Umesono, K. V., Giguere, V., Glass, C. K., Rosenfeld, M. G., and Evans, R. M. 1988. Retinoic acid and thyroid hormone induce gene expression through a common response element. *Nature* 336:262–265.

Wattenberg, L. W. 1983. Inhibition of neoplasia by minor dietary constituents. *Cancer Res. (Suppl.)* 43:2448s–2453s.

Wattenberg, L. W. 1985. Chemoprevention of cancer. *Cancer Res.* 45:1–8.

Wattenberg, L. W. 1987. Inhibitors of chemical carcinogenesis. *Adv. Cancer Res.* 26:197–226.

Wilhelmsson, A., Cuthill, S., Denis, M., Wikstrom, A.-C., Gustafsson, J.-A., and Poellinger, L. 1990. The specific DNA binding activity of the dioxin-receptor is modulated by the 90 kd heat shock protein. *EMBO J.* 9:69–76.

Wolf, C. R., Moll, E., Oesch, F., Buchmann, A., Kuhlmann, W. D., and Kunz, H. W. 1984. Characterization, localization and regulation of a novel phenobarbitol-inducible form of cytochrome P450, compared with three further P450-isoenzymes, NADPH P45-reductase, glutathione transferase and microsomal epoxide hydrolase. *Carcinogenesis* 5:993–1001.

2 | INDUCTION OF A UV DAMAGE-SPECIFIC DNA-BINDING PROTEIN CORRELATES WITH ENHANCED DNA REPAIR IN PRIMATE CELLS

Miroslava Protić, Steven Hirschfeld, Alice P. Tsang, Mary McLenigan, Kathleen Dixon, Arthur S. Levine

Section on Viruses and Cellular Biology, National Institute of Child Health and Human Development, Bethesda, Maryland

INTRODUCTION

DNA repair is a manifold process that operates to protect a cell from permanent changes in DNA structure caused by endogenous or environmental agents. If not repaired, DNA lesions can lead to loss of genomic integrity, alteration of gene expression, and ultimately cell transformation or cell death. Our appreciation of the significance of the repair processes for human health comes from extensive studies on the human hereditary disease, xeroderma pigmentosum (XP): A high susceptibility of XP patients to sun-induced skin cancer is strongly correlated with the hypersensitivity of XP cells to UV light and their decreased capacity to repair UV-induced photoproducts (Kraemer et al., 1984).

In the past decade, many efforts to purify UV lesion-specific DNA repair proteins, and to identify and clone the corresponding genes that may be defective in XP cells, have met with limited success, mainly because of the lability of purified repair proteins, the absence of highly sensitive and specific in vitro repair assays, and the difficulty of stably integrating transfected DNA in human cells (Parrish and Lambert, 1990; Wood et al., 1988; Sibghat-Ullah et al., 1989; Hoejimakers et al., 1987, 1990). Three out of eight XP complementing genes have been successfully cloned: XP A (Tanaka et al., 1990); XP BC (Weeda et al., 1990); and XP D (Weber et al., 1990), but the exact function of their corresponding products in mammalian excision repair is still hypothetical. Recently, we and others have taken a novel approach to study the molecular mechanisms of DNA repair and to identify proteins that interact with UV damaged DNA (UV-DDB).

Virus-based expression vectors are used as a model DNA in the in vivo studies of the repair of actively transcribed genes in a nonreplicating DNA environment (Fig. 1) (Kraemer et al., 1987). These vectors encode

Present address for Dr. Kathleen Dixon is Department of Environmental Health, University of Cincinnati Medical Center, University of Cincinnati, OH 45267-0056.

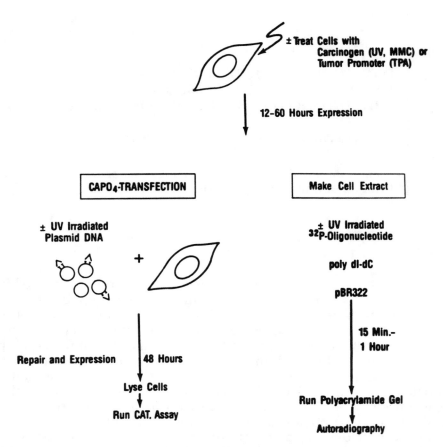

FIGURE 1. Schematic diagram of a host-cell repair assay (left column) and a damage-specific DNA-binding assay (right column). The cells were treated with carcinogens or tumor promoters and allowed to express inducible proteins. Cells were then either transfected with damaged nonreplicating expression vectors to assay for DNA repair, or used to make a cell-free extract for detection of damage-specific DNA-binding proteins. The expression vector contains the bacterial *cat* gene, the activity of which is monitored in cell lysates with a simple enzyme assay. Damage-specific DNA-binding proteins are detected as DNA-protein complexes retarded on a nondenaturing polyacrylamide gel. The radioactively labeled DNA substrate is first modified with a carcinogen (UV, cisplatin, etc.) to induce specific lesions, and then incubated with nuclear proteins in the presence of nonspecific DNA competitors (poly dI-dC, pBR322, etc.).

bacterial proteins, the expression of which can be easily monitored in a mammalian cell background. DNA lesions that are introduced into the vectors in vitro can be easily quantified and localized (Protić-Sabljić et al., 1986; Sibghat-Ullah et al., 1990). The repair assay is based on the observation that unrepaired lesions represent blocks to transcription by RNA polymerase, which result in the inhibition of gene expression. Thus, the repair capacity of cells can be assayed by measuring the ability of damaged templates to be transcribed in vivo. In addition, the expression vector-DNA repair assay permits treatment of the cells with DNA-damaging agents independent of treatment of the vector DNA to assess the indirect effects of these agents on the repair process.

Damage-specific DNA-binding (DDB) proteins are postulated to be a part of cellular DNA repair complexes with a role in targeting the other repair components to a particular lesion in DNA (Friedberg, 1985). Recently, a modification of the gel retardation assay has enabled detection and identification of several new DDB proteins in primate cells (Fig. 1) (Chu and Chang, 1988; Toney et al., 1989; Hirschfeld et al., 1990; Lenz et al., 1990; Abramić et al., 1991). Synthetic oligonucleotides (or short DNA fragments) modified in vitro with DNA-damaging agents have been used as substrates for binding of specific cell-extract proteins. The assay permits the study of species, tissue, and developmental expression of DDB proteins when extracts from cells of different genotypes and/or phenotypes are tested (McLenigan, Levine, and Protić, in press). In addition, extracts prepared from cells treated with various agents can be used to study the regulation and induction of DDB proteins (Hirschfeld et al., 1990; and this chapter).

Treatment of mammalian cells with DNA-damaging agents results in the appearance of a group of cellular phenomena that are phenotypically similar to the global DNA damage-recovery ("SOS") response of *Escherichia coli*: enhanced viral reactivation and mutagenesis [for reviews see Defais et al. (1983) and Elespuru (1987), Dion and Hamelin (1987)]; enhanced excision repair of UV-irradiated or cisplatin-treated transfected virus-based expression vectors (Protić et al., 1988; Sheibani et al., 1989); enhanced mutagenesis of UV-irradiated transfected shuttle vectors (Sarkar et al., 1984; Roilides et al., 1988; Dixon et al., 1989); and induction in UV-irradiated host cells of integrated viruses (polyoma: Ronai and Weinstein, 1988; SV40: Lavi and Etkin, 1981; Dinsart et al., 1985; HIV: Valerie et al., 1988). In addition, a number of transcripts (Angel et al., 1986; Fornace et al., 1988, 1989b; Ronai et al., 1988; Rosen et al., 1990; Woloschak et al., 1990; Devary et al., 1991) and cellular proteins (Miskin and Ben-Ishai, 1981; Schorpp et al., 1984; Keyse and Tyrrell, 1987; Stein et al., 1988; Kartasova et al., 1988; Lambert et al., 1989; Glazer et al., 1989; Angulo et al., 1989; Singh and Lavin, 1990; Ronai and Weinstein, 1990; Chao, 1991) were induced in mammalian cells exposed to DNA-damaging agents, heat shock, or tumor promoters. Whether some of these induced proteins participate in the DNA-damage recovery response in mammalian cells is not known.

Most of the induced proteins thus far identified (e.g., keratins, collagenase, stromolysin, major histocompatibility class I antigens, and the cellular protooncogenes *c-fos*, *c-jun*, and *c-myc*) are unlikely to affect cellular DNA repair directly, but may serve as intermediates in the signal transduction from damaged DNA to the damage-responsive genes (Kaina et al., 1989). However, some of the induced proteins are known to participate in DNA repair, such as DNA ligase (Mezzina and Nocentini, 1978; Li and Rossman, 1989) and O^6-methylguanine-methyltransferase (Lefebvre and Laval, 1986), or interact with damaged DNA (Brown-Luedi and Brown, 1989; Hirschfeld et al., 1990; Abramić et al., 1991; Chao, 1991). In yeast, a number of damage-inducible genes have been identified, some of which are involved in DNA repair or in damage tolerance (Sebastian et al., 1990). Moreover, three members of the yeast nucleotide excision repair pathway (RAD 2, RAD 4, and RAD 10) and yeast photolyase share a common decanucleotide sequence required for the damage-specific induction. Whether mammalian repair proteins also belong to the class of damage-responsive gene products remains to be determined, as does the biological significance of these inducible responses.

Our goal is to elucidate the mechanism(s) of enhanced DNA repair at the biochemical and the molecular levels, and to determine the significance of this inducible response in mammalian cells. Here we describe the detection of enhanced DNA repair in UV-irradiated primate cells, and the identification of an inducible, damage-specific DNA binding protein, which might participate in this recovery response.

MATERIALS AND METHODS

Cells and Treatment

The origins of monkey and human cell lines, as well as growth conditions, were previously described (Hirschfeld et al., 1990; Protić et al., 1988; Protić-Sabljić et al., 1986; Protić-Sabljić and Kraemer, 1985). Treatment of cells with UV light (UV), mitomycin C (MIT-C), 12-O-tetradecanoylphorbol-13-acetate (TPA), aphidicolin (APC), dactinomycin, and cycloheximide was described in Hirschfeld et al. (1990).

Plasmids and Oligonucleotides

The plasmid pSV2catSVgpt was used as the target DNA for in vivo DNA repair studies (Protić-Sabljić and Kraemer, 1985). It contains the bacterial *cat* gene, which codes for chloramphenicol acetyltransferase (CAT), placed under the SV40 early promoter to permit expression in mammalian cells. The ^{32}P end-labeled double-stranded oligonucleotide 3/4 (5'-GATCTGATTCCCCATCTCCTCAGTTTCACTTCTGCACCGCATG-3' with four bases overhanging on both the 5' end and 3' end of the upper strand) was used as a probe for in vitro DNA binding studies (Hirschfeld et al., 1990). The oligonucleotide is rich in adjacent pyrimidines in the upper strand that are substrates for the formation of pyrimidine dimers by UV. Oligonucleotides 3/4, 23/24 (identical to 3/4 except that a single -TTT- was substituted with -GAC-), and the double-stranded *c-fos* UV-responsive element (*fos*-URE; -329/-296; 5'-CTTTACACAGGATGTCCATATT-AGGACATCTGCG-3') were used as binding competitors. Special features of these model DNA substrates and their in vitro modification by UV light were described in Protić-Sabljić and Kraemer (1985), Stein et al. (1989), and Hirschfeld et al. (1990).

DNA Repair and DNA-Binding Assays

The capacity of mammalian cells to repair plasmid DNA was measured by transfecting cells with pSV2catSVgpt that had been UV-damaged in vitro (Protić-Sabljić and Kraemer, 1985) and assaying CAT activity in cell lysates (Neumann et al., 1987). The presence of UV-DDB proteins was determined by incubating UV-irradiated oligonucleotide 3/4 with nuclear extracts, and examining the appearance of specific protein-DNA complexes after resolution of the reaction mixtures on nondenaturing polyacrylamide gels (Hirschfeld et al., 1990) (Fig. 1).

CAT and DNA-binding activities were standardized per amount of cell extract protein. Relative activities were calculated as a ratio of treated versus mock-treated values.

RESULTS

Repair of UV-Damaged Plasmid DNA Is Enhanced in Cells Pretreated with UV but Not TPA

We have previously shown that mammalian cells exposed to DNA-damaging agents exhibit a higher capacity to repair subsequently transfected UV-damaged plasmid DNA (Protić et al., 1988). To examine whether such an effect could also be observed by treating the cells with agents that do not damage DNA, we pretreated cells with the tumor promoter TPA and measured the repair of UV-damaged plasmid. We found that monkey cells pretreated with TPA show a similar capacity to express UV-damaged plasmid DNA as mock-treated cells (Fig.2). In contrast, cells treated with UV, or UV plus TPA, repair twice as many UV-induced lesions, similar to our earlier observations. Recently, Herrlich et al. observed that a number of cellular genes, some of which may be involved in DNA repair or damage tolerance, are induced after treatment with both UV and TPA (reviewed in Kaina et al., 1989). These authors have also suggested that these cellular responses are mediated through the activation of c-fos, because fos-URE and the element required for TPA-dependent regulation of gene expression are identical. Our results imply that with the treatment conditions reported by Herrlich et al. (Schorpp et al., 1984), TPA cannot induce a cellular response that is manifested as enhanced repair of plasmid DNA. It will be of interest to learn whether cells deficient in c-fos protein still show enhanced repair after UV irradiation.

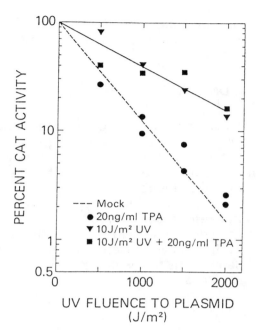

FIGURE 2. Inactivation of CAT expression in CV-1 cells either mock-treated or treated with UV, UV + TPA, or TPA only. Cells were either treated or not, and 24 h later transfected with 10 μg/dish of pSV2catSVgpt as described in Protić et al. (1988). Data are from one representative experiment. Curves were drawn by visual fitting.

Induction of a UV Damage-Specific DNA-Binding Protein in Cells Pretreated with UV, MIT-C, or APC, but Not TPA or After Serum Starvation

We postulated that mammalian cells, when pretreated with agents that induce enhanced DNA repair, have higher levels of proteins that specifically interact with damaged DNA, and that may be responsible for this enhancement. To test this hypothesis we prepared cell-free extracts from cells treated in various ways and tested the extracts in a gel retardation assay for the presence of proteins that bind to UV-irradiated DNA. As we have previously shown, monkey cells have a constitutive UV-DDB protein that binds with high affinity to UV-irradiated dsDNA (Hirschfeld et al., 1990). Extracts prepared from UV-, MIT-C-, or APC-treated cells show enhanced levels of UV-DDB protein (Figs. 3 and 4). This enhancement in UV-DDB protein appears to be the result of de novo protein synthesis, because no protein induction was detected in UV-irradiated cells treated with the transcription inhibitors dactinomycin and cyclohex-imide (Hirschfeld et al., 1990). In contrast, TPA treatment, or serum starvation, had no effect on the level of UV-DDB protein. Thus, it appears that agents that directly damage cellular DNA (UV and MIT-C), or directly inhibit DNA replication (APC), are inducers of UV-DDB protein. These results correlate with our findings on enhanced repair in the same primate cells. Although a direct involvement of UV-DDB protein in enhanced repair of UV-damaged DNA has yet to be demonstrated, it appears likely that both cellular responses, enhanced repair and enhanced UV-DDB protein synthesis, may be invoked by the same induction pathway in pretreated cells. At present, we have no evidence as to other members of this induction pathway. Our studies show, however, that *fos* may not be one of the intermediates, because no enhancement of UV repair or UV-DDB protein was detected in TPA-treated cells.

It has been postulated that UV radiation induces the posttranslational modification of a transcription factor that binds to the *fos*-URE, inducing *fos* transcription soon after UV irradiation and in the absence of any protein synthesis (Kaina et al., 1989). Although our binding probe has no

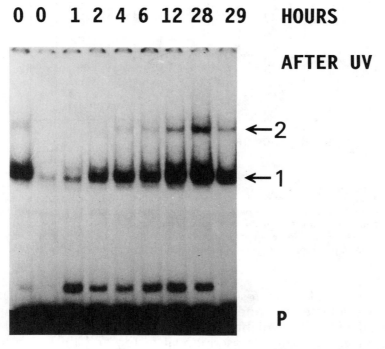

FIGURE 3. Time course of induction of UV-DDB protein in UV-irradiated CV-1 cells. Mobility shift gel with ^{32}P-labeled, UV-irradiated (9 KJ/m^2; 254 nm) oligonucleotide 3/4, and 5 μg of cell extract proteins prepared from mock-treated cells (first and last lane) or cells treated with 12 J/m^2 of 254 nm UV. Binding assay was carried out with a 100-fold molar excess of unirradiated, unlabeled 3/4 DNA. Bands 1 and 2, UV lesion specific; P, free probe. [The band that is visible on the gel in UV-treated cells, just above the probe, is not reproducible with different extract preparations; it also appears in untreated cells and its intensity cannot be correlated with a particular treatment or cell type. See also Hirschfeld et al. (1990).]

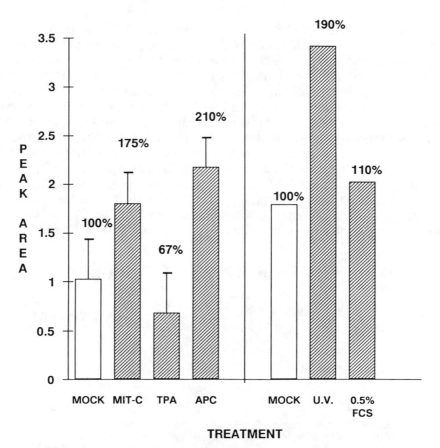

FIGURE 4. Effect of treatment of CV-1 cells on the induction of UV-DDB protein. Results of densitometer scanning (expressed as a percentage of peak areas relative to control) of the autoradiographs of mobility shift gels of the damage-specific band 1 in extracts prepared from various treated cells. (A) Results are means ± standard deviations (error bars) from two experiments. (B) Results are from one representative experiment. Conditions for the binding assay were the same as in Fig. 3.

sequence homology to *fos*-URE, and the induction of UV-DDB by UV cannot be detected in protein synthesis-inhibited cells, it is possible that UV-modified 3/4 DNA might assume the conformation required for binding of an activated *fos* transcriptional factor, and thus give a positive result in our binding assay for UV-DDB protein. However, with undamaged or UV-damaged oligonucleotides 3/4 as probes, we found that specific binding could be detected only if the DNA probes were UV damaged. Moreover, specific UV-DNA/protein complexes could be inhibited by adding a 100-fold molar excess of UV-irradiated oligonucleotides 3/4, 23/24, or *fos*-URE to the binding mixture. Finally, the same undamaged oligonucleotides had no effect on specific binding (Fig. 5). Because the three DNA competitors differ in their sequence, it is unlikely that UV modification of a particular sequence plays a role in specific binding. In addition, unirradiated *fos*-URE DNA did not compete with UV-damaged 3/4 for UV-DDB protein, and thus it is unlikely that UV-DDB protein is identical to activated *fos* transcriptional factor.

Varying Induction of UV-DDB Protein in Primate Cells with Different Phenotypes

If UV-DDB protein is involved in the repair of UV-damaged DNA, one can expect that some mammalian cells that are DNA repair-deficient should show either decreased or absent UV-DDB activity, or a defect in the regulation of UV-DDB protein expression. A number of primate cell lines were tested for the presence of the constitutive UV-DDB protein and its induction by UV (Table 1). All human primary skin fibroblasts and fetal fibroblasts show both the constitutive and the UV-induced UV-DDB protein. Cells from two (out of 10 tested) XP E patients do not show either the constitutive or the induced UV-DDB protein. Transformation of skin fibroblasts with simian virus 40 (Protić-Sabljić et al., 1986) or viral recombinant DNA (pSV7, Wood et al., 1987; Protić, unpublished) makes these cells resistant to UV-DDB induction by UV; we were unable to detect enhanced levels of UV-DDB protein 24 or 48 h after UV (Table 1, and Protić, unpublished). Similarly, monkey COS cells (CV-1 cells stably transformed with simian virus 40) have lost the ability to induce UV-DDB

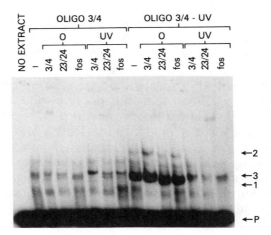

FIGURE 5. Binding of UV-DDB protein to UV-irradiated DNA in the presence of different competitors. Mobility shift gel with unirradiated (oligo 3/4) or UV-irradiated (oligo 3/4-UV) labeled DNA, and 5 μg of CV-1 cell extract proteins in the absence (−) or presence of 100-fold molar excess of unirradiated (O) or UV-irradiated (UV) unlabeled oligonucleotides 3/4, 23/24, or *fos*-URE. Bands 1 (strong) and 2, UV damage specific; band 3, nonspecific; P, free probes.

TABLE 1. Presence and Inducibility of the UV-DDB Protein in Primate Cells

Cells	UV-DDB protein	
	Constitutive	Induced[a]
Human		
Normal		
Fetal fibroblast	+	+
Primary skin fibroblast	+	+
SV40-transformed skin fibroblast	+	−
pSV7-transformed skin fibroblast	+	−
Ad-transformed embryonic kidney	+	−
Xeroderma pigmentosum		
Primary skin fibroblast XP C	+	−
Primary skin fibroblast XP E[b]	±	±
pSV7-transformed skin fibroblast XP E[c]	−	−
SV40-transformed skin fibroblast XP A	+	−
SV40-transformed skin fibroblast XP D	+	−
Monkey		
Kidney CV-1	+	+
SV40-transformed kidney (COS)	+	±

[a]Cells were irradiated with either 12 J/m^2 UV (normal), 2 J/m^2 UV (XP A and D), or 6 J/m^2 UV (XP C and E) 48 h before nuclear extract preparation.

[b]Two (one European patient, GM2415, and one Japanese patient, XP82T0) out of 10 XPE patients tested do not show UV-DDB activity. XPE cells, except GM 2415, were a generous gift from S. Kondo, Tokyo Medical and Dental University.

[c]Cells from GM2415 patient.

protein to the levels found in the parental cell line. From these experiments several interesting observations emerged: (1) primary and transformed DNA repair-deficient cells from some XP E patients are lacking both the constitutive and the induced UV-DDB activity; (2) primary cells of DNA repair-deficient XP group C, and transformed XP groups A and D, show constitutive DDB protein, but do not show induced levels of this protein 48 h after UV; and (3) virus-transformed repair-proficient cells have partially or fully lost the ability to induce UV-DDB protein.

Our interpretation of these findings remains hypothetical. XP group E cells that are lacking UV-DDB activity either do not make UV-DDB protein at all, or the protein is mutated in a DNA-binding site. Screening of XPE cell extracts with UV-DDB protein-specific antibodies should distinguish between protein deficiency and alteration.

The lack of induction of UV-DDB protein in some primary XP cell lines also can be interpreted as an inability of these XP cells to recover after UV irradiation. Although we have used two- to sixfold lower UV doses to treat XP cells than normal cells, we found that the kinetics of UV-DDB protein recovery after UV are much slower in XP cells, and reflect the degree of repair deficiency of a particular XP group (Protić, unpublished). However, UV-DDB activity in UV-irradiated XP cells might also show inducibility with longer incubation times.

The effect of viral transformation on UV-DDB protein induction by UV is intriguing. One possibility is that transformed cells have lost the ability to respond to UV through the inactivation by viral proteins of one or more intermediates in a signal transduction pathway. The other possibility is that cells have aged during the process of transformation, and have consequently lost inducing potential as a result of the shutoff of programmed gene(s). Our preliminary data on the absence of induction of the UV-DDB protein in human primary cell cultures at high passage numbers support the latter interpretation (Protić, unpublished).

DISCUSSION

We have shown that mammalian cells in culture respond to DNA damage by enhanced synthesis of a UV damage-specific DNA-binding protein, and by an associated increase in the capacity for DNA repair (Protić et al., 1988; Hirschfeld et al., 1990; and this review). These cellular responses are reminiscent of the recovery processes in *E. coli* termed "SOS," whereby a large number of genes, some of which are involved in DNA repair, are induced after DNA damage (Witkin, 1976; Walker, 1985). Although the magnitude of the enhanced DNA repair is small in comparison to induction responses in bacteria, the biological consequences of this inducible response might be significant for the mammalian cell. With either animal viruses or virus-based shuttle vectors as "probes," the enhanced reactivation of UV-irradiated replicating DNA is accompanied by mutagenesis in normal mammalian cells (for a review see Defais et al., 1983; Dixon et al., 1989; Dion and Hamelin, 1987). These observations support the possible existence of an error-prone mode of DNA repair in UV-irradiated mammalian cells. Because enhanced reactivation is usually measured using replicating viral DNA, it is not known whether the enhanced repair of the nonreplicating *cat* plasmid is also error-prone. Enhanced repair, similar to enhanced viral reactivation, can be demonstrated only at later times after UV, which suggests the requirement for de novo protein synthesis. DNA repair that occurs immediately after UV, and presumably is carried out by constitutive proteins, is error-free. Therefore, we can anticipate that cells with higher levels of the constitutive repair proteins are at lower risk for the induction of mutations, because the majority of lesions, in particular those in actively transcribed genes (Hanawalt et al., 1989), would be removed before error-prone repair can occur. For cells with low levels of constitutive DNA repair, like *E. coli*, the induced repair enables them to survive, even at the price of introducing mutations. For mammalian cells, the error-prone repair, which may result in the induction of the transformed phenotype, certainly would not be beneficial for the organism as a whole but might, nonetheless, be beneficial for the cell itself. It is possible that the presence of antioncogenes in mammalian cells serves to buffer the potential for a transformed cell phenotype, which would otherwise be the frequent consequence of error-prone repair.

Is the induction of the UV-DDB protein responsible for enhanced repair in UV-treated primate cells? A UV-DDB protein binds with high affinity to UV-damaged double-stranded oligonucleotide DNA (Hirschfeld et al., 1990; Abramić et al., 1991). The protein does not recognize oxidative lesions (Hirschfeld et al., 1990) or some bulky lesions (cyclobutane pyrimidine dimers, Hirschfeld et al., 1990; acetylaminofluorene adducts, Abramić, McLenigan, and Protić, unpublished; cisplatin crosslinks, Chu and Chang, 1988) in double-stranded DNA. Cells that have two- to threefold induced levels of UV-DDB protein also show about twofold enhanced repair of UV-damaged plasmid DNA. Chu and Chang (1990) have reported recently that a UV-DDB protein that binds to UV-damaged DNA and is lacking in some XP E cells, and thus might be identical to our UV-DDB protein, is increased in human tumor cell lines selected for resistance to cisplatin. Cisplatin-resistant cell lines also show enhanced repair of a cisplatin-damaged *cat* expression vector, and higher relative survival after cisplatin or UV treatment than parental cell lines (Sheibani et al., 1989; Chao et al., 1990; Chu and Chang, 1990). These studies strongly suggest that repair-proficient primate cells with induced levels of UV-DDB protein can repair damaged DNA with a greater efficiency. This is similar to the increased resistance of rat hepatoma cells to *N*-methyl-*N'*-nitro-*N*-nitrosoguanidine (MNNG) after various DNA-damaging treatments with a concomitant increase in the level of O^6-methylguanine-DNA-methyltransferase (Laval and Laval, 1984; Laval, 1985; Lefebvre and Laval, 1986) (Table 2). However, the mechanism of removal of methylated bases in DNA (when the methyl group from the O^6 position of guanine is transferred to a cysteine residue within the O^6-methylguanine-DNA-methyltransferase and results in the concomitant irreversible inactivation of the enzyme and restoration of the guanine moiety in the DNA) differs from the excision-gap filling-ligation process assumed for the repair of bulky lesions in mammalian cells. The induction of such a repair mechanism is associated with increased, rather than decreased, cell resistance to mutagenic effects of alkylating agents (Laval and Laval, 1984).

Although a large number of studies have shown that there might be an inducible DNA repair in mammalian cells (reviewed in Defais et al., 1983; Sarasin, 1985; Rossman and Klein, 1985; Elespuru, 1987; Denhardt and Kowalski, 1988), scant data are available on the actual mediators of these responses and their mechanisms. For enzymes that act at the later stage of excision repair, such as polymerases alpha, delta, and epsilon (PCNA-independent polymerase delta), there is no evidence for induced protein activity after UV irradiation or other DNA-damaging treatments.

TABLE 2. Induction by Physical or Chemical Agents of Proteins Involved in DNA Repair in Mammalian Cells

Protein	Inducing agent	Fold increase	Reference
DNA ligase	UV	2	Mezzina and Nocentini (1978)
	MIT-C	2	Mezzina et al. (1982)
	Acetoxyacetylaminofluorene	2	Mezzina et al. (1982)
	MNU	2.5	Li and Rossman (1989)
DNA polymerase beta	Cisplatin	3–5	Kraker and Moore (1988)
O^6-Methylguanine-DNA-methyltransferase	UV	2.5	Lefebvre and Laval (1986)
	γ-rays	5	Lefebvre and Laval (1986)
	Heat	2	Lefebvre and Laval (1986)
	Cisplatin	4	Lefebvre and Laval (1986)
	Bleomycin	2	Lefebvre and Laval (1986)
	2-Methyl-9-hydroxyellipticinium	5	Lefebvre and Laval (1986)
	MNNG	3	Laval (1985)
UV-DDB[a]	UV	2–6	Hirschfeld et al. (1990)
	MIT-C	2	Hirschfeld et al. (1990)
	APC	2	Hirschfeld et al. (1990)
	Cisplatin	2.7–4.3	Chu and Chang (1990)
	Cisplatin	3–4	Chao et al. (1990)

[a]UV-DDB protein was included in this list, although we do not yet have direct proof that it participates in DNA repair.

The activity of DNA polymerase beta is apparently elevated three- to fivefold in a cisplatin-resistant P388 murine leukemia cell line (Kraker and Moore, 1988) (Table 2). The same cell line has shown enhanced levels of DNA repair synthesis and an associated increase in the amount of proliferating cell nuclear antigen (PCNA) bound to DNA after cisplatin treatment (Haneda et al., 1991). Interestingly, the two- to fourfold induction of both polymerase beta promoter activity and mRNA in Chinese hamster ovary cells, treated with agents that produce single-base damages, is not accompanied by an increase in the level of beta-polymerase (Fornace et al., 1989). Posttranscriptional steps are likely to play a regulatory role in expression of the beta-polymerase gene after certain types of DNA damage. DNA ligase is enhanced twofold in monkey kidney cells after UV (Mezzina and Nocentini, 1978) or chemical carcinogen treatment (Mezzina et al., 1982), and 2.5-fold in Chinese hamster cells after N-methyl-N-nitrosourea (MNU) treatment (Li and Rossman, 1989) (Table 2).

Our studies show that the signal for induction of UV-DDB protein is either damaged DNA or inhibition of cellular DNA replication. This finding is not surprising, because similar responses have been observed in E. coli (Walker, 1985). It would be of interest, however, to examine the effect of other DNA-damaging agents that induce different mechanisms of repair, for example, alkylating agents or gamma-irradiation. The induction of yeast photolyase, an enzyme that specifically and exclusively photorepairs pyrimidine dimers in UV-damaged DNA, is triggered by various DNA-damaging agents, such as UV, methylmethanesulfonate, and MNNG (Sebastian et al., 1990). Thus, for some DNA repair proteins, the induction signal is not the generation of their specific substrate but rather a more general metabolic response to DNA damage. O^6-Methylguanine-DNA-methyltransferase may be another example of such a protein (Table 2).

In conclusion, we have presented evidence that enhancement in DNA repair capacity and the induction of a UV-DDB protein occur in mammalian cells after treatment with DNA-damaging agents. A high affinity for UV-damaged DNA, and the absence of specific binding in at least some repair-deficient XP E patients, suggest a role for the UV-DDB protein in DNA repair. Studies are in progress to determine the structure of the UV-DDB protein, its cellular function, and mechanisms of its regulation in untreated and UV-treated primate cells.

REFERENCES

Abramić, M., Levine, A. S., and Protić, M. 1991. Purification of an ultraviolet-inducible, damage-specific DNA-binding protein from primate cells. *J. Biol. Chem.* 266:22493–22500.

Angel, P., Poting, A., Mallick, U., Rahmsdorf, H. J., Schorpp, M., and Herrlich, P. 1986. Induction of metallothionein and other mRNA species by carcinogens and tumor promoters in primary human skin fibroblasts. *Mol. Cell. Biol.* 6:1760–1766.

Angulo, J. F., Moreau, P. L., Maunoury, R., Laporte, J., Hill, A. M., Bertolatti, R., and Devoret, R. 1989. KIN, a mammalian nuclear protein immunologically related to *E. coli* Rec A protein. *Mutat. Res.* 217:123–134.

Brown-Luedi, M. L., and Brown, T. C. 1989. Two proteins of 220 kDa and 230 kDa bind to UV-damaged SV40 minichromosomes in irradiated monkey kidney cells. *Mutat. Res.* 227:227–231.

Chao, C. C.-K. 1991. Potential negative regulation of damage-recognition proteins in cisplatin-resistant HeLa cells in response to DNA damage. *Mutat. Res.* 264:59–66.

Chao, C. C.-K., Lee, Y.-L., and Lin-Chao, S. 1990. Phenotypic reversion of cisplatin resistance in human cells accompanies reduced host cell reactivation of damaged plasmid. *Biochem. Biophys. Res. Commun.* 170:851–859.

Chu, G., and Chang, E. 1988. Xeroderma pigmentosum group E cells lack a nuclear factor that binds to damaged DNA. *Science* 242:564–567.

Chu, G., and Chang, E. 1990. Cisplatin-resistant cells express increased levels of a factor that recognizes damaged DNA. *Proc. Natl. Acad. Sci. USA* 87:3324–3327.

Defais, M. J., Hanawalt, P. C., and Sarasin, A. 1983. Viral probes for DNA repair. *Adv. Radiat. Biol.* 10:1–37.

Denhardt, D. T., and Kowalski, J. 1988. Is there induced DNA repair in mammalian cells? *BioEssays* 9:70–72.

Devary, Y., Gottlieb, R. A., Lau, L. F., and Karin, M. 1991. Rapid and preferential activation of the c-jun gene during the mammalian UV response. *Mol. Cell. Biol.* 11:2804–2811.

Dinsart, C., Cornelis, J. J., Decaesstecker, M., van der Lubbe, J., van der Eb, A. J., and Rommelaere, J. 1985. Differential effect of ultraviolet light on the induction of simian virus 40 and a cellular mutator phenotype in transformed mammalian cells. *Mutat. Res.* 151:9–14.

Dion, M., and Hamelin, C. 1987. Relationship between enhanced reactivation and mutagenesis of u.v.-irradiated human cytomegalovirus in normal human cells. *EMBO J.* 6:397–399.

Dixon, K., Roilides, E., Hauser, J., and Levine, A. S. 1989. Studies on direct and indirect effects of DNA damage on mutagenesis in monkey cells using an SV40-based shuttle vector. *Mutat. Res.* 220:73–82.

Elespuru, R. K. 1987. Inducible responses to DNA damage in bacteria and mammalian cells. *Environ. Mol. Mutagen.* 10:97–116.

Fornace, A. J., Jr., Alamo, I., Jr., and Hollander, M. C. 1988. DNA damage-inducible transcripts in mammalian cells. *Proc. Natl. Acad. Sci. USA* 85:8800–8804.

Fornace, A. J., Jr., Zmudzka, B., Hollander, C., and Wilson, S. 1989. Induction of β-polymerase mRNA by DNA-damaging agents in Chinese hamster ovary cells. *Mol. Cell. Biol.* 9:851–853.

Friedberg, E. C. 1985. *DNA Repair.* San Francisco: W. H. Freeman.

Glazer, P. M., Greggio, N. A., Metherall, J. E., and Summers, W. C. 1989. UV-induced DNA binding proteins in human cells. *Proc. Natl. Acad. Sci. USA* 86:1163–1167.

Hanawalt, P. C., Mellon, I., Scicchitano, D., and Spivak, G. 1989. Relationships between DNA repair and transcription in defined DNA sequences in mammalian cells. In *DNA Repair Mechanisms and Their Biological Implications in Mammalian Cells,* eds. M. W. Lambert and J. Laval, pp. 325–337. New York: Plenum Press.

Hirschfeld, S., Levine, A. S., Ozato, K., and Protić M. 1990. A constitutive damage-specific DNA-binding protein is synthesized at higher levels in UV-irradiated primate cells. *Mol. Cell. Biol.* 10:2041–2048.

Hoeijmakers, J. H. J., Odijk, H., and Westerveld, A. 1987. Differences between rodent and human cell lines in the amount of integrated DNA after transfection. *Exp. Cell Res.* 169:111–119.

Hoeijmakers, J. H. J., Eker, A. P. M., Wood, R. D., and Robins, P. 1990. Use of in vivo and in vitro assays for the characterization of mammalian excision repair and isolation of repair proteins. *Mutat. Res.* 236:223–238.

Kaina, B., Stein, B., Schonthal, A., Rahmsdorf, H. J., Ponta, H., and Herrlich, P. 1989. An update of the mammalian UV response: Gene regulation and induction of a protective function. In *DNA Repair Mechanisms and Their Biological Implications in Mammalian Cells,* eds. M. W. Lambert and J. Laval, pp. 149–165. New York: Plenum Press.

Kartasova, T., Ponec, M., and van de Putte, P. 1988. Induction of proteins and mRNA after UV irradiation of human epidermal keratinocytes. *Exp. Cell Res.* 174:421–432.

Keyse, S. M., and Tyrrell, R. M. 1987. Both near ultraviolet radiation and the oxidising agent hydrogen peroxide induce a 32-kDa stress protein in normal human skin fibroblasts. *J. Biol. Chem.* 262:14821–14825.

Kraemer, K. H., Lee, M. M., and Scotto, J. 1984. DNA repair protects against cutaneous and internal neoplasia: Evidence from xeroderma pigmentosum. *Carcinogenesis* 5:511–514.

Kraemer, K. H., Protić-Sabljić, M., Bredberg, A., and Seidman, M. M. 1987. Plasmid vectors for study of DNA repair and mutagenesis. *Current Problems in Dermatology,* ed. H. Honigsmann, vol. 17, pp. 166–181. Basel: S. Karger.

Kraker, A. J., and Moore, C. W. 1988. Elevated DNA polymerase beta activity in a cis-diamminedichloroplatinum (II) resistant P388 murine leukemia cell line. *Cancer Lett.* 38:307–314.

Lambert, M. E., Ronai, Z. A., Weinstein, I. B., and Garrels, J. I. 1989. Enhancement of major histocompatibility class I protein synthesis by DNA damage in cultured human fibroblasts and keratinocytes. *Mol. Cell. Biol.* 9:847–850.

Laval, F. 1985. Repair of methylated based in mammalian cells during adaptive response to alkylating agents. *Biochimie* 67:361–364.

Laval, F., and Laval, J. 1984. Adaptive response in mammalian cells: Crossreactivity of different pretreatments on cytotoxicity as compared to mutagenicity. *Proc. Natl. Acad. Sci. USA* 81:1062–1066.

Lavi, S., and Etkin, S. 1981. Carcinogen mediated induction of SV40 DNA synthesis in SV40 transformed Chinese hamster embryo cells. *Carcinogenesis* 2:417–423.

Lefebvre, P., and Laval, F. 1986. Enhancement of O^6-methylguanine-DNA-methyltransferase activity induced by various treatments in mammalian cells. *Cancer Res.* 46:5701–5705.

Lenz, J., Okenquist, S. A., LoSardo, J. E., Hamilton, K. K., and Doetsch, P. W. 1990. Identification of a mammalian nuclear factor and human cDNA-encoded proteins that recognize DNA containing apurinic sites. *Proc. Natl. Acad. Sci. USA* 87:3396–3400.

Li, J.-H., and Rossman, T. G. 1989. Inhibition of DNA ligase activity by arsenite: A possible mechanism of its comutagenesis. *Mol. Toxicol.* 2:1–9.

Mezzina, M., and Nocentini, S. 1978. DNA ligase activity in UV-irradiated monkey kidney cells. *Nucleic Acids Res.* 5:4317–4328.

Mezzina, M., Nocentini, S., and Sarasin, A. 1982. DNA ligase activity in carcinogen-treated human fibroblasts. *Biochimie.* 64:743–748.

Miskin, R., and Ben-Ishai, R. 1981. Induction of plasminogen activator by UV light in normal and xeroderma pigmentosum fibroblasts. *Proc. Natl. Acad. Sci. USA* 78:6236–6240.

Neumann, J. R., Morency, C. A., and Russian, K. O. 1987. A novel rapid assay for chloramphenicol acetyltransferase gene expression. *BioTechniques* 5:444–447.

Parrish, D. D., and Lambert, M. W. 1990. Chromatin-associated DNA endonucleases from xeroderma pigmentosum cells are defective in interaction with damaged nucleosomal DNA. *Mutat. Res.* 235:65–80.

Protić, M., Roilides, M., Levine, A. S., and Dixon, K. 1988. Enhancement of DNA repair capacity of mammalian cells by carcinogen treatment. *Somat. Cell Mol. Genet.* 14:351–357.

Protić-Sabljić, M., and Kraemer, K. H. 1985. One pyrimidine dimer inactivates expression of a transfected gene in xeroderma pigmentosum cells. *Proc. Natl. Acad. Sci. USA* 82:6622–6626.

Protić-Sabljić, M., Tuteja, N., Munson, P. J., Hauser, J., Kraemer, K. H., and Dixon, K. 1986. UV light-induced cyclobutane pyrimidine dimers are mutagenic in mammalian cells. *Mol. Cell. Biol.* 6:3349–3356.

Roilides, E., Munson, P. J., Levine, A. S., and Dixon, K. 1988. Use of a Simian virus 40-based shuttle vector to analyze enhanced mutagenesis in mitomycin C-treated monkey cells. *Mol. Cell. Biol.* 8:3943–3946.

Ronai, Z. A., Okin, E., and Weinstein,, I. B. 1988. Ultraviolet light induces the expression of oncogenes in rat fibroblast and human keratinocyte cells. *Oncogene* 2:201–204.

Ronai, Z. A., and Weinstein, I. B. 1988. Identification of a UV-induced *trans*-acting protein that stimulates polyomavirus DNA replication. *J. Virol.* 62:1057–1060.

Ronai, Z. A., and Weinstein, I. B. 1990. Identification of ultraviolet-inducible proteins that bind to a TGACAACA sequence in the polyoma virus regulatory region. *Cancer Res.* 50:5374–5381.

Rosen, C. F., Gajic, D., and Drucker, D. J. 1990. Ultraviolet radiation induction of ornithine decarboxylase in rat keratinocytes. *Cancer Res.* 50:2631–2635.

Rossman, T. G., and Klein, C. B. 1985. Mammalian SOS system: A case of misplaced analogies. *Cancer Invest.* 3:175–187.

Sarasin, A. 1985. SOS response in mammalian cells. *Cancer Invest.* 3:163–174.

Sarkar, S., Dasgupta, U. B., and Summers, W. C. 1984. Error-prone mutagenesis detected in mammalian cells by a shuttle vector containing the *sup F* gene of *Escherichia coli. Mol. Cell. Biol.* 4:233–237.

Schorpp, M., Mallick, U., Rahmsdorf, H. J., and Herrlich, P. 1984. UV-induced extracellular factors from human fibroblasts communicates the UV response to nonirradiated cells. *Cell* 37:861–868.

Sebastian, J., Kraus, B., and Sancar, G. B. 1990. Expression of the yeast *PHR1* gene is induced by DNA-damaging agents. *Mol. Cell. Biol.* 9:4630–4637.

Sheibani, N., Jennerwein, M. M., and Eastman, A. 1989. DNA repair in cells sensitive and resistant to cis-diamminedichloroplatinum (II): Host cell reactivation of damaged plasmid DNA. *Biochemistry* 28:3120–3124.

Sibghat-Ullah, Husain, I., Carlton, W., and Sancar, A. 1989. Human nucleotide excision repair in vitro: Repair of pyrimidine dimers, psoralen and cisplatin adducts by HeLa cell-free extract. *Nucl. Acid Res.* 17:4471–4484.

Sibghat-Ullah, Sancar, A., and Hearst, J. E. 1990. The repair patch of *E. coli* (A)BC excinuclease. *Nucleic Acid Res.* 18:5051–5053.

Singh, S. P., and Lavin, M. 1990. DNA-binding protein activated by gamma radiation in human cells. *Mol. Cell. Biol.* 10:5279–5285.

Stein, B., Rahmsdorf, H. J., Schonthal, A., Buscher, M., Ponta, H., and Herrlich, P. 1988. The UV induced signal transduction pathway to specific genes. In *The Mechanisms and Consequences of DNA Damage Processing*, eds. E. Friedberg and P. Hanawalt, pp. 557–570. New York: Alan R. Liss.

Stein, B., Rahmsdorf, H. J., Steffen, A., Litfin, M., and Herrlich, P. 1989. UV-induced DNA damage is an intermediate step in UV-induced expression of human immunodeficiency virus type 1, collagenase, *c-fos*, and metallothionein. *Mol. Cell. Biol.* 9:5169–5181.

Tanaka, K., Miura, N., Satokata, I., Miyamoto, I., Yoshida, M. C., Satoh, Y., Kondo, S., Yasui, A., Okayama, H., and Okada, Y. 1990. Analysis of a human DNA excision repair gene involved in group A xeroderma pigmentosum and containing a zinc-finger domain. *Nature* 348:73–76.

Toney, J. H., Donahue, B. A., Kellett, P. J., Bruhn, S. L., Essigmann, J. M., and Lippard, S. J. 1989. Isolation of cDNA encoding a human protein that binds selectively to DNA modified by the anticancer drug *cis*-diamminedichloroplatinum (II). *Proc. Natl. Acad. Sci. USA* 86:8328–8332.

Valerie, K., Delers, A., Bruck, C., Thiriart, C., Rosenberg, H., Debouck, C., and Rosenberg, M. 1988. Activation of human immunodeficiency virus type 1 by DNA damage in human cells. *Nature* 333:78–81.

Walker, G. C. 1985. Inducible DNA repair systems. *Annu. Rev. Biochem.* 54:425–457.

Weber, C. A., Salazar, E. P., Stewart, S. A., and Thompson, L. H. 1990. ERCC-2: cDNA cloning and molecular characterization of a human nucleotide excision repair gene with high homology to yeast *RAD3*. *EMBO J.* 9:1437–1447.

Weeda, G., van Ham, R. C. A., Masurel, R., Westerveld, A., Odijk, H., de Wit, J., Bootsma, D., van der Eb, A. J., and Hoeijmakers, J. H. J. 1990. Molecular cloning and biological characterization of the human excision repair gene ERCC-3. *Mol. Cell. Biol.* 10:2570–2581.

Witkin, E. M. 1976. Ultraviolet mutagenesis and inducible DNA repair in *Escherichia coli*. *Bacteriol. Rev.* 40:869–907.

Woloschak, G. E., Chang-Liu, C.-M., and Shearin-Jones, P. 1990. Regulation of protein kinase C by ionizing radiation. *Cancer Res.* 50:3963–3967.

Wood, C. M., Timme, T. L., Hurt, M. M., Brinkley, B. R., Ledbetter, D. H., and Moses, R. E. 1987. Transformation of DNA repair-deficient human diploid fibroblasts with a Simian virus 40 plasmid. *Exp. Cell Res.* 169:543–553.

Wood, R. D., Robins, P., and Lindahl, T. 1988. Complementation of the xeroderma pigmentosum DNA repair defect in cell-free extracts. *Cell* 53:97–106.

3 | REPAIR OF ALKYLATED BASES IN MAMMALIAN CELLS: MODULATION IN RESPONSE TO DNA-DAMAGING AGENTS

Françoise Laval

Group "Radiochimie de l'ADN," INSERM, Institut Gustave Roussy, Villejuif, France

INTRODUCTION

In order to keep the integrity of their genetic material, bacteria and mammalian cells possess a variety of DNA repair mechanisms (reviewed by Laval and Laval, 1980; Lindahl, 1982). Besides the constitutive repair enzymes, bacterial cells respond to DNA damage by inducing different groups of genes, organized as functional units called regulons (Gottesman, 1984), which code for proteins involved in DNA repair, mutagenesis, and recombination (reviewed by Walker, 1984). These regulons respond to a variety of stresses; among them are the SOS (*lexA*-controlled) regulon (Walker, 1984), the heat shock (*htpr*-controlled) regulon (Neidhart et al., 1985) and the oxidation stress (*oxyR*-controlled) regulon (Christman et al., 1985).

Similar inducible responses exist in mammalian cells. Ultraviolet (UV) reactivation of viruses has been shown to be mutagenic and inducible (Sarasin and Hanawalt, 1978; DasGupta and Summers, 1978). Heat-shock proteins, which increase the cellular thermotolerance, are synthesized after various stresses (Carper et al., 1987), including treatment with the chloroethylnitrosoureas (Kroes et al., 1991). An enhanced resistance to oxidative damage has also been shown in human lymphocytes treated with low doses of ionizing radiations (Wolff et al., 1988) and in two different cell lines treated with repeated nontoxic doses of oxygen species (Laval, 1988), although the involvement of DNA repair in these increased resistances has not yet been demonstrated.

When cells are treated with alkylating agents, various lesions are produced in DNA, including modified bases, sugars, and phosphotriesters (Singer and Grunberger, 1983). Among them, the 3-methyladenine (3-meAde) residues, which block DNA synthesis (Boiteux et al., 1984), and O^6-methylguanine (O^6-meGua) residues, which change base pairing

The author thanks Dr. T. O'Connor for the human ANPG cDNA, Dr. S. Mitra for the human transferase cDNA, and Mrs. Monique Charra for performing the densitometry measurements. The technical assistance of Mrs. Martine Letourneur was greatly appreciated. This work was supported by grants from Institut National de la Santé et de la Recherche Médicale and from Association pour la Recherche sur le Cancer (Villejuif).

(Swann, 1990), may lead to lethal or mutagenic events. As a response to such damage, *Escherichia coli* has both constitutive and inducible repair proteins. The constitutive repair of 3-meAde residues is performed by 3-meAde-DNA-glycosylase I (Tag I) and to a lesser extent by 3-meAde-DNA-glycosylase II (Tag II) coded for by the *tag* A and *alk* A genes, respectively (reviewed by Sakumi and Sekiguchi, 1990). The repair of O^6-meGua residues is performed by two O^6-meGua-DNA-methyltransferase (transferase) proteins, a 39-kD inducible protein and a 19-kD constitutive protein, coded for by the *ada* and *ogt* genes, respectively (for review see Takano et al., 1991). Exposure of *E. coli* cells to nontoxic doses of an alkylating agent triggers the adaptive response in which the Ada and Tag II protein activities increase, resulting in a higher resistance of the bacteria to the toxic and mutagenic effects of alkylating drugs (Samson and Cairns, 1977; Jeggo et al., 1977). This adaptive response differs from the regulons described earlier because it is specifically triggered by the alkylation of the *ada* gene product (Shevell et al., 1991).

As alkylating drugs are potent mutagens and carcinogens (Saffhill et al., 1985) and are also used in cancer therapy (D'Incalci et al., 1988; Pegg, 1990), many attempts have been made to discover whether a similar adaptive response exists in mammalian cells (reviewed by Frosina and Abbondandolo, 1985). Here we present data showing that the repair of alkylation damage can be increased in various mammalian cell lines by a process that is different from the *E. coli* adaptive response. This repair is increased in cells treated with DNA-damaging agents, and these treatments enhance the transcription of the 3-meAde glycosylase and of the transferase genes.

MATERIALS AND METHODS

Cell Culture

The H4 cells (derived from a rat hepatoma) and LICH cells (derived from a human hepatoma) (Laval and Little, 1977) were grown in Dulbecco's medium supplemented with 5% fetal calf serum, 5% horse serum, penicillin (50 units/ml) and streptomycin (50 μg/ml) in a humidified 5% CO_2 atmosphere. The doubling times were about 15 h and 24 h for H4 and LICH cells, respectively.

For survival measurements, the cells were subcultured in an appropriate number and grown for 14 d until appearance of clones. Restriction enzymes were introduced in the cells by electroporation: 0.5 ml of cell suspension (10^7 cells/ml) in 272 mM sucrose, 7 mM Na$_2$HPO$_4$, 1 mM MgCl$_2$, pH 7.4, was electroporated (400 V, 25 μF) at 0°C in the presence of the different enzymes, in a 0.4-cm cuvette (Biorad Gene Pulser apparatus). After electroporation, the cells were maintained for 10 min at 0°C, then diluted and grown in normal medium.

Irradiations and Drug Treatment

γ-Rays were delivered by a ^{60}Co γ-ray source at a dose rate of 1.0 Gy/min. Ultraviolet irradiation was carried out with a General Electric 254 nm germicidal lamp, at a fluence rate of 10 J/m^2/s. cis-Dichloroammine platinum(II) (cis-DDP) (Roger Bellon, Paris, France) and 2-methyl-9-hydroxyellipticinium (Institut Pasteur, Paris, France) were dissolved in water, then diluted to the appropriate concentrations in the culture medium. Other drugs were directly diluted in culture medium.

Determination of the O^6-Methylguanine-DNA-methyltransferase Activity

This activity was measured as already described (Lefebvre and Laval, 1986). Briefly, the cells were disrupted by sonication and increasing amounts of cell extracts were incubated for 20 min at 37°C with [^{3}H]-N-methyl-N-nitrosourea ([^{3}H]MNU)-treated DNA. The substrate was then acid-hydrolyzed and the remaining O^6-methylguanine was measured after separation by high-performance liquid chromatography (HPLC).

Determination of the N^3-Methyladenine-DNA-glycosylase Activity

Cell extracts were incubated for 30 min at 37°C with [^{3}H]dimethyl sulfate ([^{3}H]DMS)-treated DNA, prepared as already described (Laval, 1985). The samples were then precipitated by adding 12.5 μg calf thymus DNA, 100 μg bovine serum albumin, and ethanol at 0°C. After centrifugation, the supernatants were supplemented with authentic methylated purines and analyzed by HPLC.

RNA Purification and Northern Blot Analysis

Total RNA was extracted according to the method of Chirgwin et al. (1979): 30 μg RNA was separated by electrophoresis on formaldehyde-containing agarose gels, then transferred to nitrocellulose by capillarity. The probes were the Xba I/Bam H1 fragment of the plasmid containing the rat transferase cDNA (Rahden-Staron and Laval, 1991) and the Xba 1/Hind III fragment of the APDG plasmid expressing the rat 3-meAde glycosylase (O'Connor and Laval, 1990). They were labeled by nick translation and had a specific activity of about 4×10^8 cpm/μg. Hybridization was run as described by Maniatis et al. (1989). Hybridization was quantified by densitometry measurements of autoradiographs (Laval, 1991).

SDS-Polyacrylamide Gel Electrophoresis

The H4 cells extracts were incubated with [^{3}H]MNU-treated DNA (corresponding to 2000 cpm O^6-meGua) for 20 min at 37°C; then the samples were run in 12.5% polyacrylamide gels containing sodium dodecyl sulfate (SDS) (Laemmli, 1970). The gels were treated with Enhance (NEN Chemical Co.), dried, and exposed with intensifying screens.

RESULTS

We have previously shown (Lefebvre and Laval, 1986) that the transferase activity was increased in different cell lines treated with a single dose of various physical or chemical DNA-damaging agents (γ-rays, UV light, alkylating drugs, ellipticine, bleomycin, etc.). Figure 1 shows the number of transferase molecules per cell in two hepatoma cell lines, of rat (H4 cells) or human (LICH cells) origin. These cells, which have a different constitutive transferase activity, were treated with equitoxic doses of the various agents. Each treatment resulted in 30 ± 5% survival (measured by cloning efficiency), and a time length corresponding to about three cell doublings separated the treatment and the transferase activity determination (48 and 72 h for H4 and LICH cells, respectively). The results show that the number of transferase molecules per cell in-

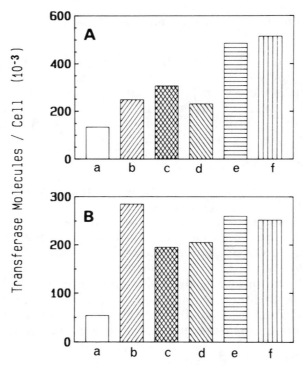

FIGURE 1. Number of transferase molecules in H4 and LICH cells treated with various DNA-damaging agents. The number of transferase molecules was calculated as already described (Lefebvre and Laval, 1986) from control cells (a), or from cells treated with γ-rays (b), or incubated for 1 h with paraquat (c), MMS (d), cis-DDP (e), or NMHE (f). A, LICH cells; B, H4 cells.

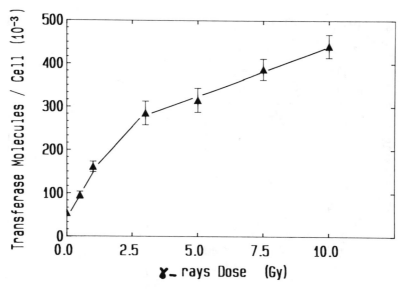

FIGURE 2. Number of transferase molecules in γ-irradiated H4 cells. Exponentially growing cells were irradiated with increasing γ-ray doses, then grown for 48 h prior to the transferase activity determination.

creases after treatments that form different types of damage in the cellular DNA, for example, after oxidative stress produced by ionizing radiations or paraquat, different alkylation damage produced by MMS or *cis*-DDP, or drug intercalation (*N*-methyl-9-hydroxyellipticinium, NMHE). For a single inducing agent, the level of induction (two- to fivefold) varies with the cell line tested. Using different cell lines, we have shown that treatment with a single dose of the various inducing agents results in a maximum enhancement after about three cell doublings; then the number of transferase molecules begins to decrease to reach the control value (Laval, 1990). However, when the cells are treated with repeated doses (e.g., when they receive a γ-ray dose of 1 Gy every 48 h), the transferase activity is increased as long as the treatment is delivered to the cells, although the magnitude of the enhancement is similar whether the cells receive a single or several doses of the inducing agent (data not shown).

The transferase activity increases with the dose of inducing agent delivered to the cells, then reaches a plateau. Figure 2 shows the number of transferase molecules per H4 cell 48 h after irradiation with increasing γ-ray doses. A similar dose-dependent increase was observed in cells treated with chemical compounds (e.g., *cis*-DDP, NMHE) (data not shown).

The 3-meAde-DNA-glycosylase activity is also increased by these treatments, but to a lesser extent than the transferase one (Fig. 3): 11, 27,

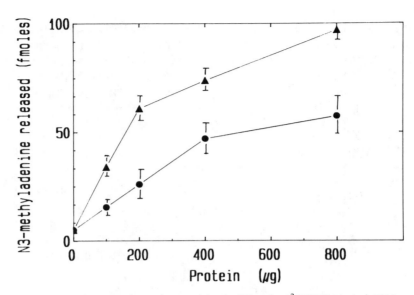

FIGURE 3. The 3-meAde-DNA-glycosylase activity in H4 cells. [³H]DMS-treated DNA was incubated with extracts of control (●) or γ-irradiated (3 Gy) (▲) cells. The glycosylase activity was measured 48 h after the irradiation.

and 20 fmol of 3-meAde residues were released from [³H]DMS-treated DNA by cellular extracts containing 100 μg of proteins from control, γ-irradiated (3 Gy), or *cis*-DDP-treated (5 μM) H4 cells, respectively.

De novo protein synthesis is required to increase these repair activities, as their enhancement is not observed when the cells are treated with the inducing agent then grown in the presence of cycloheximide (1 μg/ml for 18 h) (Frosina and Laval, 1987). In the case of the transferase, the newly synthesized molecules have the same molecular weight as the constitutive one (Fig. 4), and like the constitutive molecules (Brent et al., 1988), in our experimental conditions, they only remove the O^6-meGua residues from alkylated DNA and show activity neither on O^4-methylthymine nor on phosphotriesters.

Different sets of evidence suggest the role of DNA damage in the induction of the two repair proteins. Their activities are not modified when the cells are treated with compounds that do not directly interact with the cellular DNA (e.g., uncouplers of oxidative phosphorylation, metabolic inhibitors). The inducing treatments produce DNA breaks, either directly (e.g., γ-rays) or during the repair of the damage they have produced in the cellular DNA (e.g., UV light, alkylating agents). The role of DNA damage in this induction is also supported by the fact that a greater enhancement of the transferase activity is obtained when the cells are treated with the inducing agent than when they are grown in the presence of a poly(ADP-ribose) synthesis inhibitor (Lefebvre and

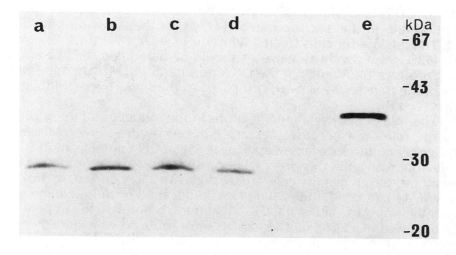

FIGURE 4. SDS-polyacrylamide gel electrophoresis (SDS-PAGE) of H4 cells extracts incubated with [³H]MNU-treated DNA. Extracts were prepared from control cells (*a*), or 48 h after γ-irradiation (3 Gy) (*b*) or treatment with *cis*-DDP (5 μM) (*c*) or NMHE (2.5 μg/ml) (*d*). They were incubated with [³H]MNU-treated DNA, then subjected to SDS-PAGE. *Escherichia coli* extracts were used as control (*e*).

Laval, 1989), which could act either by interacting with DNA repair or by modifying the DNA structure (reviewed by Gaal and Pearson, 1985). Furthermore, the transferase activity is enhanced when restriction enzymes are introduced in the cells by electroporation (Table 1) under conditions that produce DNA damage (detected by nucleoid sedimentation) and/or

TABLE 1. Number of Transferase Molecules in H4 Cells 48 Hours After Treatment with Restriction Enzymes

Cell treatment	Number of transferase molecules/cell
None	54,000 ± 6000
Electroporation without enzyme	59,000 ± 5800
Alu I (20 U)	145,000 ± 2600
Pvu II (20 U)	216,700 ± 1400
Eco RI (40 U)	102,600 ± 2300

Note. The cells were electroporated in the presence of the enzymes as described in Materials and Methods, grown for 48 h, and then the transferase activity was determined (Lefebvre and Laval, 1986).

chromatid breaks (data not shown). These restriction enzymes had similar toxicities in the electroporated H4 cells, as the surviving fraction was 30, 34, and 36% for cells treated with Alu I (20 U), PVU II (20 U), and Eco RI (40 U), respectively. However, it should be noted that the two enzymes that produce blunt ends in the DNA (Alu I and PVU II) are better inducers of the transferase activity, compared to the sticky-ends-producing enzyme Eco RI.

The role of cell-cycle modifications in the induction of these proteins can be ruled out because they are induced by the same treatments in stationary cultures (Lefebvre and Laval, 1986). DNA amplification is probably not involved in this process, because (1) there is no induction of the repair proteins when the cells are treated with a low dose of hydroxyurea (1 mM, 4 h), in conditions where DNA replication is inhibited and subsequently resumes, although this treatment is known to amplify DNA (Johnston et al., 1986), (2) the karyotype of induced cells does not show the structures (minichromosomes or HSR zones) frequently associated with DNA amplification (Schimke, 1988), and (3) Southern blot analysis of the genomic DNA from control or treated H4 cells does not show detectable amplification.

In order to elucidate the molecular mechanism of this induction, we have isolated the cDNA, encoding either the 3-meAde glycosylase (O'Connor and Laval, 1990) or the transferase (Rahden-Staron and Laval, 1991) from H4 cells. Using these probes, experiments were designed to measure the abundance of the RNA transcripts in the treated cells. When total RNA was extracted from control or treated cells and then hybridized with the cDNA, an increase of the transcripts was observed 24 and 48 h after the cell treatments. This increase was transient, as it disappeared after 96 h (Table 2), and this time-dependent increase followed the same kinetics as that of the repair-protein activity. A similar increase of the RNA transcripts was measured in LICH cells treated with various DNA-damaging agents, using as probes either the cDNA expressing the human 3-meAde-DNA-glycosylase (ANPG cDNA) (O'Connor and Laval, 1991) or the human transferase (Tano et al., 1990) (data not shown).

Although it is transient, this enhancement of the repair proteins has biological properties. When H4 cells are pretreated with DNA-damaging agents and then incubated with [^{3}H]MNU, the rate of removal of O^6-meGua and 3-meAde residues is faster and more complete in the pretreated cells compared to control cultures (Laval, 1990). Numerous data have shown that the cell sensitivity to the toxic effect of the chloroethylnitrosoureas was related to the number of transferase molecules per cell (reviewed by D'Incalci et al., 1988; Pegg, 1990). Therefore, one can expect that H4 cells pretreated with DNA-damaging agents would become more resistant to the killing effect of these alkylating compounds. Results summarized in Table 3 represent the D_0 doses (dose corresponding to a 37% survival) for H4 cells treated with BCNU and show that the BCNU-

TABLE 2. Abundance of the mRNA Transcripts in H4 Cells After Different Treatments

Cell treatment	Time after treatment	β-Actin mRNA	Transferase mRNA	3-meAde-glycosylase mRNA
None		1	1	1
γ-Rays (3 Gy)	24 h	1.0	3.8	1.8
	48 h	1.0	4.9	2.8
	96 h	0.9	1.05	1.2
cis-DDP (5 μM)	24 h	1.0	3.1	1.7
	48 h	1.2	4.5	2.7
	96 h	1.3	1.5	0.9
NMHE (2.5 μg/ml)	24 h	1.1	5.5	1.9
	48 h	0.8	5.1	2.5
	96 h	1.2	1.6	1.1

Note. H4 cells were treated with the various agents and the RNA extracted at different times after the treatments. The amounts of β-actin, transferase, and 3-meAde-DNA-glycosylase mRNA were determined by densitometry measurement after slot blot hybridization of total RNA (20 μg) with the appropriate probes. The values represent the abundance of the mRNA in the treated cells, compared to control cultures.

resistance of these cells increases after the pretreatments that enhance the transferase activity (Habraken and Laval, 1991).

DISCUSSION

The results presented in this chapter show that two proteins involved in the repair of alkylated bases are increased in cells treated by a variety of DNA-damaging agents. It should be stressed that among the different cell lines tested, this enhancement seems to occur preferentially in transformed cells (Laval, 1990). This process is different from the adaptive

TABLE 3. Influence of Various Pretreatments on the Number of Transferase Molecules per H4 Cell and on the D_0 for BCNU-Treated Cells

Cell pretreatment	Number of transferase molecules/cell[a]	D_0 for BCNU-treated cells $(\mu g/ml)$[b]
None	54,000 ± 6000	12.1 ± 1.2
γ-Rays (3 Gy)	285,000 ± 27,600	22.3 ± 2.5
cis-DDP (5 μM)	207,000 ± 12,200	18.0 ± 1.7
NMHE (2.5 μg/ml)	252,000 ± 6000	21.8 ± 2.2

[a]These values were calculated from the linear part of the curves obtained by incubating increasing amounts of cell extracts with [³H]MNU-treated DNA (Lefebvre and Laval, 1986) and are the mean value ± SD of three separate experiments.
[b]Calculated from three different survival curves (Habraken and Laval, 1991).

response to alkylating drugs that occurs in *E. coli*, because it is not triggered only by an alkylating compound, but actually by a variety of physical or chemical agents, which all interact with the cellular DNA. These
inducing treatments produce various types of DNA damage, which are
further repaired by different pathways, and this suggests that the type of
inducing damage is rather broad and might favor the role of a structural
modification of the DNA molecule in this induction.

Different cDNA clones encoding DNA-damage inducible transcripts
have been isolated from rodent cells (Fornace et al., 1989a) and from
human cells (Ben Ishai et al., 1990). Although most of the proteins encoded by these transcripts have not yet been characterized, it has been
shown that DNA damage induced the expression of collagenase, metallothionein, c-*fos* (Stein et al., 1989), β-polymerase (Fornace et al., 1989b),
and DNA-binding protein (Hirschfeld et al., 1990). However, induction of
these transcripts appears soon (between 2 and 8 h) after the cell treatments, whereas the induction of the 3-meAde-glycosylase and of the
transferase transcripts appears later (between 24 and 48 h) after the treatments. This suggests that there may be different regulatory pathways
controlling gene expression in response to DNA damage, resulting in
different kinetics of induction.

Although the enhancement of the 3-meAde-glycosylase and of the
transferase activities is transient, it was of interest to ask whether the
activation of these two genes had biological properties. In fact, our
results have shown that it has a protective value against the toxic effect
of a chloroethylnitrosourea used in cancer chemotherapy [1,3-bis(2-
chloroethyl-1-nitrosourea, BCNU] (Habraken and Laval, 1991). This might
have particular relevance in oncology during the use of combined treatments with radio- and chemotherapy.

REFERENCES

Ben-Ishai, R., Scharf, R., Sharon, R., and Kapten, I. 1990. A human cellular sequence implicated in
    trk oncogene activation is DNA damage inducible. *Proc. Natl. Acad. Sci. USA* 87:6039–6043.
Boiteux, S., Huisman, O., and Laval, J. 1984. 3-Methyladenine residues in DNA induce *sfiA* in
    Escherichia coli. *EMBO J.* 3:2569–2573.
Brent, T. P., Dolan, M. E., Fraenkel-Conrat, H., Hall, J., Karran, P., Laval, F., Margison, G. P., Monte-
    sano, R., Pegg, A. E., Potter, P. M., Singer, B., Swenberg, J. A., and Yarosh, D. B. 1988. Repair of
    O-alkylpyrimidines in mammalian cells. *Proc. Natl. Acad. Sci. USA* 85:1759–1762.
Carper, S. W., Duffy, J. J., and Gerner, E. W. 1987. Heat shock proteins in thermotolerance and other
    cellular processes. *Cancer Res.* 47:5249–5255.
Chirgwin, J., Przybyla, A., McDonald, R., and Rutter, W. J. 1979. Isolation of biologically active
    ribonucleic acid from sources enriched in ribonuclease. *Biochemistry* 18:5294–5299.
Christman, M. F., Morgan, R. W., Jacobson, F. S., and Ames, B. N. 1985. Positive control of a regulon
    for defenses against oxidation stress and some heat-shock proteins in *Salmonella typhimu-
    rium*. *Cell* 41:753–762.
DasGupta, U. B., and Summers, W. C. 1978. Ultraviolet reactivation of herpes simplex virus is
    mutagenic and inducible in mammalian cells. *Proc. Natl. Acad. Sci. USA* 75:2378–2381.

D'Incalci, M., Citti, L., Taverna, P., and Catapano, C. V. 1988. Importance of the DNA repair enzyme O^6-alkylguanine alkyltransferase (AT) in cancer chemotherapy. *Cancer Treat. Rev.* 15:279–292.

Fornace, A. J., Nebert, D. W., Hollander, C., Fornace, A. J., Papathanasiou, M., Fargnoli, J., and Holbrook, N. J. 1989a. Mammalian genes coordinately regulated by growth arrest signals and DNA-damaging agents. *Mol. Cell. Biol.* 9:4196–4203.

Fornace, A. J., Zmudzka, B., Hollander, C., and Wilson, S. H. 1989b. Induction of β-polymerase mRNA by DNA-damaging agents in Chinese hamster ovary cells. *Mol. Cell. Biol.* 9:851–853.

Frosina, G., and Abbondandolo, A. 1985. The current evidence for an adaptive response to alkylating agents in mammalian cells, with special reference to experiments with in vitro cell culture. *Mutat. Res.* 154:85–100.

Frosina, G., and Laval, F. 1987. The O^6-methylguanine-DNA-methyltransferase activity of rat hepatoma cells is increased after a single exposure to alkylating agents. *Carcinogenesis* 8:91–95.

Gaal, J. C., and Pearson, C. K. 1985. Eukaryotic nuclear ADP-ribosylation reactions. *Biochem. J.* 230:1–18.

Gottesman, S. 1984. Bacterial regulation: Global regulatory networks. *Annu. Rev. Genet.* 18:415–441.

Habraken, Y., and Laval, F. 1991. Enhancement of 1,3-bis(2-chloroethyl)-1-nitrosourea resistance by γ-irradiation or drug pretreatment in rat hepatoma cells. *Cancer Res.* 52:1217–1220.

Hirschfeld, S., Levine, A. S., Ozato, K., and Protic, M. 1990. A constitutive damage-specific DNA-binding protein is synthesized at higher levels in UV-irradiated primate cells. *Mol. Cell. Biol.* 10:2041–2048.

Jeggo, P., Defais, M., Samson, L., and Schendel, P. 1977. An adaptive response of *E. coli* to low levels of alkylating agent: Comparison with previously characterized DNA repair pathways. *Mol. Gen. Genet.* 157:1–9.

Johnston, R. N., Fedder, J., Hill, A. B., Sherwood, S. W., and Schimke, R. T. 1986. Transient inhibition of DNA synthesis results in increased dihydrofolate reductase synthesis and subsequent increased DNA content per cell. *Mol. Cell. Biol.* 10:3373–3381.

Kroes, R. A., Abravaya, K., Seidenfeld, J., and Morimoto, R. I. 1991. Selective activation of human heat shock gene transcription by nitrosourea antitumor drugs mediated by isocyanate-induced damage and activation of heat shock transcription factor. *Proc. Natl. Acad. Sci. USA* 88:4825–4829.

Laemmli, P. M. 1970. Cleavage of structural proteins during the assembly of the heat of bacteriophage T4. *Nature* 227:680–685.

Laval, F. 1985. Repair of methylated bases in mammalian cells during adaptive response to alkylating agents. *Biochimie* 67:361–364.

Laval, F. 1988. Pretreatment with oxygen species increases the resistance of mammalian cells to hydrogen peroxide and γ-rays. *Mutat. Res.* 201:73–79.

Laval, F. 1990. Induction of proteins involved in the repair of alkylated bases in mammalian cells by DNA-damaging agents. *Mutat. Res.* 233:211–218.

Laval, F. 1991. Increase of O^6-methylguanine-DNA-methyltransferase and N^3-methyladenine glycosylase RNA transcripts in rat hepatoma cells treated with DNA-damaging agents. *Biochem. Biophys. Res. Commun.* 176:1086–1092.

Laval, F., and Laval, J. 1980. Enzymology of DNA repair. In *Molecular and Cellular Aspects of Carcinogen Screening Tests,* eds. R. Motesano, H. Bartsch, and L. Tomatis, pp. 55–73. Lyon: IARC.

Laval, F., and Little, J. B. 1977. Enhancement of survival of X-irradiated mammalian cells by the uncoupler of oxidative phosphorylation, *m*-chloro carbonyl cyanide phenylhydrazone. *Radiat. Res.* 71:571–578.

Lefebvre, P., and Laval, F. 1986. Enhancement of O^6-methylguanine-DNA-methyltransferase activity induced by various treatments in mammalian cells. *Cancer Res.* 46:5701–5705.

Lefebvre, P., and Laval, F. 1989. Potentiation of *N*-methyl-*N*'-nitro-*N*-nitrosoguanidine-induced-O^6-methylguanine-DNA-methyltransferase activity in a rat hepatoma cell line by poly(ADP-ribose) synthesis inhibitors. *Biochem. Biophys. Res. Commun.* 163:599–604.

Lindahl, T. 1982. DNA repair enzymes. *Annu. Rev. Biochem.* 51:61–87.

Maniatis, T., Fritsch, E. F., and Sambrock, J. 1989. *Molecular Cloning: A Laboratory Manual.* Cold Spring Harbor, N.Y.: Cold Spring Harbor Laboratory Press.

Neidhardt, F. C., Van Bogelen, R. A., and Vaughn, V. 1985. The genetics and regulation of heat-shock proteins. *Annu. Rev. Genet.* 18:295–329.

O'Connor, T. R., and Laval, F. 1990. Isolation and structure of a cDNA expressing a mammalian 3-methyladenine-DNA glycosylase. *EMBO J.* 9:3337–3342.

O'Connor, T. R., and Laval, J. 1991. Human cDNA expressing a functional DNA glycosylase excising 3-methyladenine and 7-methylguanine. *Biochem. Biophys. Res. Commun.* 176:1170–1177.

Pegg, A. E. 1990. Mammalian O^6-alkylguanine-DNA-alkyltransferase: Regulation and importance in response to alkylating carcinogenic and therapeutic agents. *Cancer Res.* 50:6119–6129.

Rahden-Staron, I., and Laval, F. 1991. cDNA cloning of the rat O^6-methylguanine-DNA-methyltransferase. *Biochem. Biophys. Res. Commun.,* in press.

Saffhill, R., Margison, G. P., and O'Connor, P. J. 1985. Mechanisms of carcinogenesis induced by alkylating agents. *Biochem. Biophys. Acta* 823:111–145.

Sakumi, K., and Sekiguchi, M. 1990. Structures and functions of DNA glycosylases. *Mutat. Res.* 236:161–172.

Samson, L., and Cairns, J. 1977. A new pathway for DNA repair in *Escherichia coli. Nature (Lond.)* 267:281–283.

Sarasin, A. R., and Hanawalt, P. C. 1978. Carcinogens enhance survival of UV-irradiated simian virus 40 in treated monkey kidney cells; Introduction of a recovery pathway? *Proc. Natl. Acad. Sci. USA* 75:346–350.

Schimke, R. T. 1988. Gene amplification in cultured cells. *J. Biol. Chem.* 263:5989–5992.

Shevell, D. E., Friedman, B. M., and Walker, G. C. 1990. Resistance to alkylation damage in *Escherichia coli*: Role of the Ada protein in induction of the adaptive response. *Mutat. Res.* 233:53–72.

Singer, B., and Grunberger, D. 1983. *Molecular Biology of Mutagens and Carcinogens.* New York: Plenum Press.

Stein, B., Rahmsdorf, H. J., Steffen, A., Liftin, M., and Herrlich, P. 1989. UV-induced DNA damage is an intermediate step in the UV-induced expression of human immunodeficiency virus type I, collagenase, c-fos, and metallothionein. *Mol. Cell. Biol.* 9:5169–5181.

Swann, P. F. 1990. Why do O^6-alkylguanine and O^4-alkylthymine miscode? The relationship between the structure of DNA containing O^6-alkylguanine and O^4-alkythymine and the mutagenic properties of these bases. *Mutat. Res.* 223:81–94.

Takano, K., Nakamura, T., and Sekiguchi, M. 1991. Role of two types of O^6-methylguanine-DNA-methyltransferases in DNA repair. *Mutat. Res.* 254:37–44.

Tano, K., Shiota, S., Collier, J., Foote, R. D., and Mitra, S. 1990. Isolation and structural characterization of a cDNA clone encoding the human DNA repair protein for O^6-alkylguanine. *Proc. Natl. Acad. Sci. USA* 87:686–690.

Walker, G. 1984. Mutagenesis and inducible responses to deoxyribonucleic acid damage in *Escherichia coli. Microbiol. Rev.* 18:60–93.

Wolff, S., Afzal, V., Wiencke, J. K., Olivieri, G., and Michaeli, A. 1988. Human lymphocytes exposed to low doses of ionizing radiation become refractory to high doses of the radiation as well as to chemical mutagens that induce double-strand breaks in DNA. *Int. J. Radiat. Biol.* 53:39–48.

4 | MODIFICATIONS IN DNA LIGASE ACTIVITY DURING DNA REPAIR IN MAMMALIAN CELLS

Mauro Mezzina

Cancer Research Biology Laboratory, Department of Radiation Oncology, Stanford University Medical Center, Stanford, California

Alain Sarasin

Laboratory of Molecular Genetics, Institut de Recherches Scientifiques sur le Cancer, Villejuif, France

INTRODUCTION

The cellular pathways induced in response to various DNA-damaging treatments in mammalian cells involve multiple genes that participate in DNA repair processes. In bacteria, these treatments induce the expression of new functions (the SOS pathway) by turning on simultaneously several genes, particularly the *recA* gene, that are under the control of the same repressor, the LexA protein (Walker, 1985). By using modern techniques of genetics and molecular biology, attempts were made in the past years to identify mammalian genes involved in DNA repair processes and in cellular stress caused by DNA damages. Some genes were identified to participate to the nucleotide excision repair pathway, such as the *ERCC* (excision repair cross-complementing rodent repair-deficiency) genes (Thompson, 1989; Weber et al., 1990; Weeda, 1990) or the *gadd* (grow arrest and DNA damage inducible) genes (Fornace et al., 1989a; Luety et al., 1990). The activation of other genes has been found to be correlated to DNA-damaging treatment (Kaina et al., 1989; Ben-Ishai et al., 1990). However, no proteins encoded by the above genes have been yet purified and demonstrated to have a clear function in DNA repair process, and all proteins identified to be induced after DNA-damaging

This work was supported by grants from the Commission of the European Community (number B17-0034, Brussels, Belgium) and A.R.C. (Association pour la Recherche sur le Cancer, Villejuif, France). The authors are grateful to Dr. U. Bertazzoni (C.N.R., Pavia, Italy), Drs. R. H. Elder and J. M. Rossignol (C.N.R.S., Villejuif, France) for enzyme purifications and basic support in scientific discussion, and Drs. A. Giaccia and M. J. Brown (Stanford University, California) for helpful discussion and criticism to the manuscript.

treatments are still to be characterized in detail (Boothman et al., 1989; Glazer et al., 1989).

The general model for DNA repair process implies that breakage of DNA at the level of or near the lesion leads to lesion removal, followed by a DNA polymerization step and subsequent rejoining of the phosphodiester backbone. A DNA joining activity is thus required to accomplish these processes. However, the detailed mechanisms of DNA repair pathway are still unknown, especially in mammalian cells. All DNA ligases purified in prokaryotic and eukaryotic systems are known to catalyze covalent ligation between 5'-phosphoryl and adjacent 3'-hydroxyl ends in double-stranded DNA. Since such nicks occur during semiconservative DNA synthesis, DNA ligase is universally considered as necessarily involved in DNA replication. In bacteria (Pauling and Hamm, 1968) and in lower eukaryotes (Fabre and Roman, 1979), genetic evidences demonstrated the involvement of the same DNA ligase in both DNA replication and repair processes. In fact, temperature-sensitive mutants for ligase activity are not only unable to grow in absence of functional DNA ligase, but are also extremely sensitive to killing by DNA-damaging treatments. Furthermore, it has been shown that the ligase yeast *cdc 9* gene is induced both during S-phase and DNA repair process occurring after UV-irradiation or methylmethanesulfonate treatment (Peterson et al., 1985).

In mammalian cells, a similar genetic approach for studying the role of the enzyme during DNA replication and repair has not been possible so far since mutant cell lines defective in ligase activity are still unavailable, and the study of this enzyme at genetic level has just begun with the recent identification of the human gene encoding for DNA ligase (Barnes et al., 1990). Therefore, studies carried out so far to elucidate the role of the enzyme during DNA replication or repair were made only by using the conventional biochemical approaches to investigate the activity levels and properties of the enzyme after its purification in normal cells either during proliferation or after DNA damage. In this chapter we discuss the properties of mammalian DNA ligase isolated in different cellular systems and using different methods. The analysis of the enzyme present in normal cells and in cells after treatment with various genotoxic agents gave in the last years further information on the nature of the enzyme and on its function during the biochemical pathways elicited by DNA damage.

The biochemical analysis of mammalian DNA ligases was carried out by using different substrates of double-stranded DNA containing single-stranded nicks (5'-phosphoryl and 3'-hydroxyl termini). Such "nicked DNA" was used to mimic DNA structure during semiconservative replicative synthesis or during excision repair. Nicked DNA can be generated either by hydrolysis of double-stranded DNA with DNases or by annealing to long single-stranded homopolymers with complementary oligonucleotides (Kornberg, 1980). This type of DNA has been ubiquitously used as the substrate for the detection of ligase activity. All prokaryotic and eukaryotic DNA ligases are known to seal nicks via the formation of covalent phosphodiester bonds in the presence of ATP or NAD cofactors (Lehman, 1974).

Using these DNA templates, the enzyme purifications, carried out from different biological systems in different laboratories, showed that two enzyme forms can be isolated, according to differences in molecular weight and chromatographic mobility. Most of the works indicate that, in several animal tissues or cultured cell lines, the predominant form of the enzyme, DNA ligase I, is an enzyme of 220–130 kD as estimated by gel filtration analysis or polyacrylamide gel electrophoresis (Söderhäll and Lindhal, 1974; Teraoka and Tsukada, 1982, 1985; Tomkinson et al., 1990). This form is considered to participate in DNA replication, since its activity increases in proliferating cells (Söderhäll, 1976). The second form, DNA ligase II, is present at lower levels and appeared to be a smaller molecular weight enzyme of 85–68 kD (Söderhäll and Lindhal, 1974; Teraoka et al., 1986). The latter enzyme has been suggested to participate in DNA repair (Söderhäll, 1976; Creissen and Shall, 1982; Mezzina et al., 1982a, 1982b). These two forms of DNA ligase can be distinguished by other criteria such as the different antigenic properties (Söderhäll and Lindhal, 1975; Teraoka et al., 1986) and, more recently, the different ability to ligate an oligo(dT) annealed to a poly(rA) (Arrand et al., 1986).

However, some of the above properties of the enzyme seem to depend on the techniques used for purification and analysis. By using two different protocols for purifying rat liver enzyme, we were able to demonstrate that it was possible to obtain one or two chromatographically distinct forms of ligase, and when two forms were obtained, both were correlated to the same 130-kD polypeptide (Mezzina et al., 1987). The same result was obtained with human cells: only DNA ligase I was detected after Superose 12 FPLC elutions of crude extracts. However, after ammonium sulfate or polymin P precipitations of the same extracts, a second peak of enzyme activity corresponding to DNA ligase II was detected (Mezzina et al., 1989). Using the specificity of a poly(rA):oligo(dT) substrate to differentiate the two forms of the enzyme, results conflicting with those previously reported have been described: DNA ligase II appeared as a larger enzyme than ligase I after gel filtrations or viscosity gradients (Elder and Rossignol, 1990). Despite the large amount of work performed on the biochemical analysis of mammalian DNA ligases, the nature of the mammalian enzyme and its function during the cell cycle and DNA repair still need to be elucidated. Moreover, the basic equation of whether one or two enzymes exist in mammalian cells remains unanswered.

METHODS

Cells

African monkey kidney cells (CV-1 P) were from Dr. P. Berg's laboratory (Stanford, Calif.); diploid human normal KD fibroblasts (CRL1295 cells) were from the American Type Culture Collection (ATCC); human embryonic lung fibroblasts (HEL-1) were generously provided by Dr. R. Cassingena (Villejuif, France); SV40-transformed fibroblasts from normal donors (HGOV5) were transformed and provided by Dr. L. Daya-Grosjean (Villejuif, France); Boom's syndrome (BS) SV40-transformed fibroblasts (GM8505) were from ATCC. These cell lines were grown or processed for genotoxic treatment essentially as already described (Mezzina et al., 1982a, 1982b, 1989).

Preparation of Crude Extracts

In most experiments, cell extracts were obtained from confluent cells by sonication in the presence of buffer A, which contained 0.5 M NaCl, 20 mM Tris-HCl, pH 8.0, 1 mM EDTA, 2 mM DTT, 0.5% Triton X-100, 10 mM sodium bisulfite, 1 mM PMSF, and 2 μg/ml each of leupeptine and pepstatine. These sonicants were then centrifuged in a refrigerated Eppendorf centrifuge. Crude cell extracts were prepared from SV40-transformed human cells with buffer B, which contained all components of buffer A except that NaCl concentration was 0.1 M and Triton X-100 was omitted (Mezzina et al., 1989).

DNA Ligase Assay

The substrate for the enzyme assay was obtained by annealing 5′-^{32}P-labeled oligo(dT)$_{16}$ molecules to a poly(dA)$_{500}$ homopolymer. The ligase activity is defined as the amount of ^{32}P material converted into an alkaline phosphatase-resistant form.

Activity Gel Technique

This method permits the detection of the ligase active polypeptides in crude or partially purified enzyme fractions. Protein samples were run in denaturing conditions in polyacrylamide gels containing the ^{32}P-oligo(dT):poly(dA) substrate. After removal of sodium dodecyl sulfate (SDS) from the gel, the ligase activity occurs in situ after protein renaturation inside the gel, and polypeptides containing ligase activity are revealed after alkaline phosphatase treatment (Mezzina et al., 1984).

Partial Purification of DNA Ligase

Phosphocellulose and hydroxylapatite chromatographies were generally employed in the enzyme purifications. Crude extracts were submitted to the ammonium sulfate (AmS) precipitation at 75% of saturation.

After the first chromatography, pooled fractions were reprecipitated with AmS before a second round of chromatography.

RESULTS

Levels of Enzyme Activity in Crude Extracts or Partially Purified Fractions from CV-1 Cells and Human Fibroblasts

CV-1 cells, which are permissive to the lytic cycle of simian virus 40 (SV40), can be studied during both DNA replication and repair processes. The viral infection of CV-1 cells induces a new round of replicative DNA synthesis in confluent monolayer cultures. Parallel cultures can be treated with different DNA-damaging agents, and ligase activity is monitored during DNA repair process. By using this approach, we measured the enzyme activity in crude cell extracts and in partially purified fractions. By using this approach, we measured the enzyme activity in crude cell extracts and in partially purified fractions. When we analyzed total DNA ligase activity in CV-1 cells treated with UV light, acetoxyacetyl-aminofluorene (AAAF), mitomycin C (MMC), or iododeoxyuridine (IUdR), we found it was increased two- to threefold compared to control cells (Table 1). This increase of enzyme activity is of the same order of magnitude as that observed in replicating cells infected by SV40. When MMC- or IUdR-pretreated cells were infected with SV40, the ligase activity increased in an additive manner compared to the increase observed when genotoxic treatments or viral infection were done separately. Similar results were obtained with human diploid fibroblasts (KDN) treated with UV light, AAAF, MMC, or 4-nitroquinoline-N-oxide (4NQO). However, when these cells were treated with N-methyl-N'-nitro-N-nitrosoguanidine (MNNG) or with ethylmethanesulfonate (EMS), no increase of ligase activity was observed. The increase of enzyme activity occurring after DNA damaging was completely eliminated by the treatment of CV-1 cells with the protein synthesis inhibitor cycloheximide, suggesting that enzyme activity requires a de novo protein synthesis. In mouse livers from animals treated in vivo with hepatocarcinogens such as bromophenylbenzanthracene, the ligase activity was twice that of untreated animals in both crude extracts and partially purified fractions (Mezzina and Szafard, unpublished results). These results obtained in vivo with organs from whole animals confirm the previous data obtained with cultured cells.

When ligase activity was analyzed after partial purification through sucrose gradient centrifugation or phosphocellulose and hydroxylapatite chromatography, we observed that enzyme activity was composed of two forms in CV-1 and human cells (Mezzina et al., 1982a, 1982b, 1985). Referring to the classical criteria cited earlier (molecular size, chromatographic properties, and sedimentation coefficient), these two forms could corre-

TABLE 1. DNA Ligase Activity in Mammalian Cell Lines after Various DNA-Damaging Treatments

Cells	Controls	UV light (17.5 J/m^2)	AAAF (3 µg/ml)	MMC (5 µg/ml)	IUdR (100 µg/ml)	MMC + IUdR	SV40	MMC + SV40	IUdR + SV40
Monkey kidney cells (CV1)	5.1	10.5	9.8	10.8	14.5	15	10.5	16.5	21.6
+ cycloheximide (5µg/ml)	4.6	2.7	—	3.5	4.5	5.3	4.6	5.3	6.1

Cells	Controls	UV light	AAAF	MMC	MNNG	EMS	4NQO
Human fibroblasts (kD)	3.1	6.5	6.9	6.3	3.4	2.9	5.2
HEL-1	3		6.7	6.3			

Note. Values are the specific activity of the enzyme (units/mg proteins). Data are obtained from Mezzina et al. (1982a, 1982b).

spond to DNA ligase I and II. The sedimentation coefficients of these two forms were 5.5–7 S for ligase I and about 4 S for ligase II. During DNA repair, the 4 S form was increased in CV-1 and in human cells treated with DNA-damaging agents (Table 1), while during DNA replication in SV40-infected CV-1 cells the heavy form was increased (Mezzina et al., 1982a, 1982b). Both peaks of enzyme activity were increased in MMC- or IUdR-pretreated infected CV-1 cells (Mezzina et al., 1982a). These results obtained after partial enzyme purification confirmed the measurements of total ligase levels in the same cellular crude extracts (Table 1). By measuring the total activity recovered after gradients, we found three to four times more activity in treated cells than in controls. This increase in ligase activity could be ascribed to the light or the heavy form in cells treated with DNA-damaging agents or infected with SV40, respectively.

Activity Gel Analysis of Mammalian DNA Ligases

The activity gel technique was recently developed to identify the active polypeptides of mammalian DNA ligase in either crude extracts or purified fractions. This technique allows study of the polypeptide composition and posttranslational modifications that may occur physiologically or during enzyme purifications. This analysis showed that human DNA ligase in crude extracts from both normal (HGOV5) and BS (GM8505) fibroblasts was composed of two major polypeptides of 200 and 130 kD (Fig. 1, lanes 1 and 2). By adding higher molar salt concentrations and detergent during the preparation of crude extracts, the 200-kD polypeptide disappeared and smaller species of 90, 75, and 60 kD became visible (lanes 3 and 4). After sucrose gradient sedimentations of crude extracts from HGOV5 cells we obtained two peaks of about 7 and 4 S (Mezzina et al., 1982b). The activity of the first peak correlated with the 130-kD polypeptide, whereas the second peak correlated with a 90-kD one (lanes 5 and 6). Polypeptides of 200 and 130 kD were also detected in partially purified fresh fractions from regenerating rat livers (lane 7). Identical results have been obtained by using normal rat livers (data not shown). However, the 200-kD polypeptide disappeared in 2-d-old enzyme fraction (lane 8). All these polypeptides corresponded to ligase activity, as determined by analysis of the size of oligo(dT) substrate recovered from radioactive bands on DNA sequencing gels (Mezzina et al., 1984). The smaller band of 43 kD did not possess ligase activity by this analysis. In CV-1 cell extracts, the majority of enzyme activity was correlated to the 90-kD band (lane 9). In MMC-, IUdR-, or MMC + IUdR-treated cells, the intensity of this band was much stronger, and lower bands of 70 kD were also visible (lanes 10–12). SV40 infection of cells produced a strong increase in the intensity of the 90-kD band, but two higher molecular weight bands of 125 and 115 kD were also visible (lane 13). The intensity of the latter bands was strongly increased in SV40-

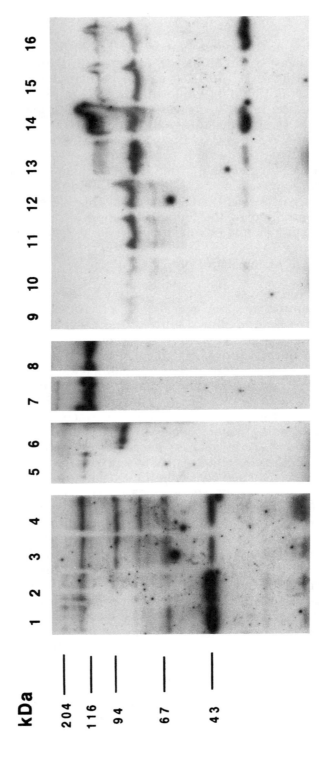

FIGURE 1. Activity gel analysis of DNA ligase from human, rat liver and monkey kidney cells: 10 μg of proteins of crude extracts from normal HGOV5 (lanes 1 and 3) and GM8505 Bloom's syndrome (lanes 2 and 4) cells were prepared by using buffer B (lanes 1 and 2) or buffer A (lanes 3 and 4) as described in Methods. Lanes 5 and 6: 15 μl of the 7 and 4 S peak fractions, respectively from sucrose gradient sedimentations of crude extracts as in lane 3. Lanes 7 and 8: 2 μg of proteins of enzyme fraction from regenerating rat liver after phosphocellulose and hydroxylapatite fractionation for freshly prepared (lane 7) or 2-d-old fraction (lane 8) samples. Lanes 9–16: 10 μg of proteins of crude extracts from monkey kidney cells, prepared as described in Methods, corresponding to untreated cells (lane 9), MMC-, IUdR-, and MMC + IUdR-treated cells (lanes 10–12), SV40-infected cells (lane 13), and MMC-, IUdR-, and IUdR+MMC-pretreated SV40 infected cells (lanes 14–16).

58

infected MMC-pretreated cells and, to a lesser extent, in SV40-infected IUdR or MMC + IUdR-pretreated cells (lanes 14–16).

Therefore, it appears that the increase of ligase activity observed in crude extracts (Table 1) is confirmed by the activity gel analysis. The treatment of cells with DNA-damaging agents preferentially promotes the increase of lower molecular weight polypeptides, whereas after SV40 infection all polypeptides were increased. Similar results were obtained when DNA ligase was purified after phosphocellulose and hydroxylapatite chromatographies from CV-1 control, MMC-treated, or SV40-infected cells. This purification procedure allowed the separation of two enzyme peaks. When DNA ligase was purified from MMC-treated cells, two to three times more enzyme was recovered in the second peak after hydroxylapatite elution, whereas with SV40-infected cells the enzyme increase was found in the first peak. This was supported by analysis of the first and second ligase forms eluted in parallel chromatographies from control cells (Franchi, Izzo, and Mezzina, unpublished results). Activity gel analysis of these fractions revealed that DNA ligase I correlated to polypeptides of 120, 110, 70, and 58 kD, and DNA ligase II to polypeptides of 65 and 58 kD. After SV40 infection, all polypeptides in both fractions were strongly increased, whereas after MMC-treatment, the disappearance of the high-molecular-weight polypeptides in both fractions correlated with an accumulation of smaller-molecular-weight species of 65, 58, and 43 kD (Mezzina et al., 1985).

DISCUSSION

Analysis of DNA ligase activity in mammalian cells during DNA repair suggests that the increased enzyme activity results from de novo protein synthesis, since this increase was completely abolished by cycloheximide (Table 1). Therefore, the increase of total ligase activity might result from transcriptional regulation, as has been suggested for *Saccharomyces cerevisiae* (Peterson et al., 1985). Similar DNA ligase activity increase after DMS treatment of mouse cells has also been hypothesized to be due to the activation of the enzyme via the poly(ADP)ribosylation pathway during DNA repair (Creissen and Shall, 1982), although this result has never been developed further. Studies on the biochemical properties of mammalian DNA ligase of different systems strongly suggest that this activity is required for completing different DNA repair pathways. Even in plant organisms, a significant increase of ligase activity was found to be correlated to DNA repair induced after gamma irradiation of rye embryos (Elder et al., 1987). The role of DNA ligase in repair, detected with one of the standard substrates, poly(dA):oligo(dT), seems to be common in several prokaryotic and eukaryotic systems. Our data indicate that this increase of ligase activity is associated with the treatment with agents such as UV light, MMC, IUdR, AAAF, or 4NQO, but not EMS or MNNG (Table

1). All these agents produce different bulky lesions, which may be re-paired by different biochemical pathways. One can speculate that chromatin structure may also play an essential role in processing different DNA damages. In fact, human cells from genetically transmitted diseases such as xeroderma pigmentosum are unable to repair damage induced by UV light or other chemicals considered to be "UV-like," such as AAAF or 4NQO, but are proficient in the repair of ionizing-radiation-induced damage such as DNA double-stranded breaks. Cell lines derived from individuals afflicted with Fanconi's anemia or ataxia telangiectasia are extremely deficient in repairing cross-links or chromosome breaks, re-spectively, but are proficient in repairing UV or "UV-like" induced lesions (Hanawalt and Sarasin, 1986). It is possible that different biochemical pathways utilize the ligase activity, which is detectable by conventional substrate and purification procedures, but its inducibility might not be necessary in some other DNA repair patches, such as those induced after alkylating agents treatment. Similarly, DNA polymerase β, which may be involved in DNA repair (Cleaver, 1983; Dresler and Lieber-man, 1983; Yamada et al., 1985), seems to be regulated differently after different types of DNA damage. In fact, β polymerase gene transcripts were enhanced threefold after MNNG, MMS, or AAAF treatment of Chi-nese hamster cells, but not after UV irradiation (Fornace et al., 1989b).

The recent isolation of human gene encoding DNA ligase I has pro-vided important insights into the nature of this mammalian enzyme. From the nucleotide sequence of the gene, the size of the protein has been definitively established to be 102 kD. However, the size of the puri-fied DNA ligase after polyacrylamide gel electrophoresis appears differ-ent: it is 125–130 kD. This difference between predicted size and experi-mental values for molecular weight might be explained taking into account the average charge of the protein, its deduced amino acid com-position (Barnes et al., 1990; Tomkinson et al., 1990), and its secondary and tertiary structure. The active species of 200 kD, detected both after immunoassay with antiligase antibodies in mouse cells (Teraoka and Tsu-kada, 1985) and by activity gel analysis of enzyme preparations from rat livers and human cells (Fig. 1), might be explained as the result of a posttranslational modification of the enzyme. It is likely that these higher polypeptides correspond to some modified forms of newly synthesized enzyme, detected in fresh preparations, due to strong protein–protein or protein–DNA intermediate complexes or to transient chemical modifica-tions. In fact, the half-life of the 200-kD polypeptide has been found to be around 30 min, after pulse-chase experiments (Teraoka and Tsukada, 1985). In our enzyme preparations, we found that the higher-molecular-weight (200–170 kD) polypeptides have short half-lives, since they were only detected in the fresh fraction of rat liver or by modifying already existing protocols for enzyme preparation from human cells (Fig. 1). It is possible that the higher-molecular-weight polypeptides detected in our experiments could be due to strong associations between subunit forms of the enzyme or between DNA ligase and other factors. On the other hand, the enzyme modifications that occurred in monkey kidney cells after DNA-damaging treatment yielded smaller fragments (Mezzina et al., 1985). It is possible that in this case such modifications were the result of increased protease activity, which may play a role in modifying the enzyme structure. In our experiments, we did not detect any increase of activity of DNA ligase II, by using poly(rA):oligo(dT) substrate, in crude extracts or partially purified fractions from human or monkey kidney cells, after treatments with MMC or UV irradiation (Mezzina and Lisandri, unpublished results). Despite the extensive purification of such an activity from calf thymus (Teraoka et al., 1986) and rat liver (Elder and Rossignol, 1990) tissues, and the isolation of antibodies raised against calf thymus enzyme (Teraoka et al., 1986), no reports have yet demonstrated that this substrate-specific and immunologically distinct form is specifically involved in cellular functions induced after the treatment with DNA-damaging agents. In order to detect other DNA ligases in mammalian cells involved in DNA repair pathways, new purification procedures and/or enzyme substrates need to be developed, with the final aim to isolate and clone the structural gene.

REFERENCES

Arrand, J. E., Willis, A. E., Goldsmith, I., and Lindhal, T. 1986. Different substrate specificities of the two DNA ligases of mammalian cells. *J. Biol. Chem.* 261:9079–9082.

Barnes, D. E., Johnston, L. H., Kodama, K., Tomkinson, A. E., Lasko, D. D., and Lindhal, T. 1990. Human DNA ligase I cDNA: Cloning and functional expression in *Saccharomyces cerevisiae*. *Proc. Natl. Acad. Sci. USA* 87:6679–6683.

Ben-Ishai, R., Scharf, R., Sharon, R., and Kapten, I. 1990. A human cellular sequence implicated in *trk* oncogene activation is DNA damage inducible. *Proc. Natl. Acad. Sci. USA* 87:6039–6043.

Boothman, D. A., Bouvard, I., and Hughes, E. N. 1989. Identification and characterization of x-ray-induced proteins in human cells. *Cancer Res.* 49:851–853.

Cleaver, J. E. 1983. Structure of repaired sites in human DNA synthesized in the presence of inhibitors of DNA polymerases α and β in human fibroblasts. *Biochim. Biophys. Acta* 739:301–311.

Creissen, D., and Shall, S. 1982. Regulation of DNA ligase activity by poly(ADP-ribose). *Nature* 296:271–272.

Dresler, S. L., and Lieberman, M. W. 1983. Identification of DNA polymerases involved in DNA excision repair in diploid human fibroblasts. *J. Biol. Chem.* 258:9990–9994.

Elder, R. H., Dell'Aquila, A., Mezzina, M., Sarasin, A., and Osborne, D. J. 1987. DNA ligase in repair and replication in the embryos of rye, *Secale cereale*. *Mutat. Res.* 181:61–71.

Elder, R. H., and Rossignol, J. M. 1990. DNA ligases from rat liver. Purification and partial characterization of two molecular forms. *Biochemistry* 29:6009–6017.

Fabre, F., and Roman, H. 1979. Evidence that a single DNA ligase is involved in replication and recombination in yeast. *Proc. Natl. Acad. Sci. USA* 76:4586–4588.

Fornace, A. J., Jr., Nebert, D. W., Hollander, M. C., Luety, J. D., Papathanasiou, M., Fargnoli, J., and Holbrook, N. J. 1989a. Mammalian genes coordinately regulated by grow arrest signals and DNA-damaging agents. *Mol. Cell. Biol.* 9:4196–4203.

Fornace, A. J., Jr., Zmudzka, B., Hollander, M. C., and Wilson, S. H. 1989b. Induction of β-

polymerase mRNA by DNA-damaging agents in Chinese hamster ovary cells. *Mol. Cell. Biol.* 9:851–853.

Glazer, P. M., Greggio, N. A., Metherall, J. E., and Summers, W. C. 1989. UV-induced DNA-binding proteins in human cells. *Proc. Natl. Acad. Sci. USA* 86:1163–1167.

Hanawalt, P. C., and Sarasin, A. 1986. Cancer prone hereditary diseases with DNA processing abnormalities. *Trends Genet.* 2:124–129.

Kaina, B. Ë., Stein, B., Schontal, A., Rahmsdorf, H. J., Ponta, H., and Herrlich, P. 1989. An update of the mammalian UV response: Gene regulation and induction of a protective function. In *DNA Repair Mechanisms and Their Implications in Mammalian Cells*, eds. M. W. Lambert and J. Laval, pp. 149–156. New York: Plenum Press.

Kornberg, A. 1980. *DNA Replication*. San Francisco: W. H. Freeman.

Lehman, I. R. 1974. DNA ligase: Structure, mechanism, and function. *Science* 186:790–797.

Luety, J. D., Fargnoli, J., Park, J., Fornace, A. J., Jr., and Holbrook, N. J. 1990. Isolation and characterization of the hamster *gadd 53* gene. *J. Biol. Chem.* 265:16521–16526.

Mezzina, M., Suarez, H. G., Cassingena, R., and Sarasin, A. 1982a. Increased activity of polynucleotide ligase in 5-iodo-2'-deoxyuridine and mitomycin C-pretreated simian virus 40 (SV40)-infected monkey kidney cells. *Nucleic Acids Res.* 10:5073–5084.

Mezzina, M., Nocentini, S., and Sarasin, A. 1982b. DNA ligase activity in carcinogen-treated human fibroblasts. *Biochimie* 64:743–748.

Mezzina, M., Sarasin, A., Politi, N., and Bertazzoni, U. 1984. Heterogeneity of mammalian DNA ligase detected on activity and DNA sequencing gels. *Nucleic Acids Res.* 12:5109–5122.

Mezzina, M., Franchi, E., Izzo, R., Bertazzoni, U., Rossignol, J. M., and Sarasin, A. 1985. Variation in DNA ligase structure during repair and replication processes in monkey kidney cells. *Biochem. Biophys. Res. Commun.* 132:857–863.

Mezzina, M., Rossignol, J. M., Philippe, M., Izzo, R., Bertazzoni, U., and Sarasin, A. 1987. Mammalian DNA ligase: Structure and function in rat liver tissues. *Eur. J. Biochem.* 162:325–332.

Mezzina, M., Nardelli, J., Nocentini, S., Renault, G., and Sarasin, A. 1989. DNA ligase activity in human cell lines from normal donors and Bloom's syndrome patients. *Nucleic Acids Res.* 17:3091–3106.

Pauling, C., and Hamm, L. 1968. Properties of a temperature-sensitive radiation-sensitive mutant of *Escherichia coli. Proc. Natl. Acad. Sci. USA* 60:1495–1502.

Peterson, T. A., Prakash, L., Prakash, S., Osley, M. A., and Reed, S. J. 1985. Regulation of *CDC 9*, the *Saccharomyces cerevisiae* gene that encodes DNA ligase. *Mol. Cell. Biol.* 5:226–235.

Söderhäll, S. 1976. DNA ligases during rat liver regeneration. *Nature* 260:640–642.

Söderhäll, S., and Lindhal, T. 1974. DNA ligases of eukaryotes. *FEBS Lett.* 67:1–7.

Söderhäll, S., and Lindhal, T. 1975. Mammalian DNA ligases. Serological evidences for two separate enzymes. *J. Biol. Chem.* 250:8438–8444.

Teraoka, H., and Tsukada, K. 1982. Eukaryotic DNA ligase. Purification and properties of the enzyme from bovine thymus, and immunochemical studies of the enzyme from animal tissues. *J. Biol. Chem.* 257:4758–4763.

Teraoka, H., and Tsukada, K. 1985. Biosynthesis of mammalian DNA ligase. *J. Biol. Chem.* 260:2937–2940.

Teraoka, H., Sumikawa, T., and Tsukada, K. 1986. Purification of DNA ligase II from calf thymus and preparation of rabbit antibody against calf thymus DNA ligase II. *J. Biol. Chem.* 261:6888–6892.

Thompson, L. H. 1989. Somatic cell genetic approach to dissecting mammalian DNA repair. *Environ. Mol. Mutagen.* 14:264–281.

Tomkinson, A. E., Lasko, D., Daly, G., and Lindhal, T. 1990. Mammalian DNA ligases. Catalytic domain and size of DNA ligase I. *J. Biol. Chem.* 265:12611–12617.

Walker, G. C. 1985. Inducible DNA repair systems. *Annu. Rev. Biochem.* 54:425–457.

Weber, C. A., Salazar, E. P., Stewart, S. A., and Thompson, L. 1990. ERCC2:cDNA cloning and molecular characterization of a human nucleotide excision repair gene with high homology to yeast *RAD3. EMBO J.* 9:1437–1447.

Weeda, G., van Ham, R. C. A., Vermeulen, W., Bootsma, D., van der Eb, A., and Hoeijmakers, J. H. J.

1990. A presumed DNA helicase encoded by *ERCC-3* is involved in the human repair disorders xeroderma pigmentosum and Cockayne's syndrome. *Cell* 62:777–781.

Yamada, K., Hanaoka, F., and Yamada, M. 1985. Effect of aphidicolin and/or 2′ or 3′-dideoxythymidine on DNA repair induced in HeLa cells by four types of DNA-damaging agents. *J. Biol. Chem.* 260:10412–10417.

5 | CELL-CYCLE AND DNA-DAMAGE REGULATION OF RIBONUCLEOTIDE REDUCTASE IN *Saccharomyces cerevisiae*

Stephen J. Elledge, Zheng Zhou

Department of Biochemistry and Institute for Molecular Genetics, Baylor College of Medicine, Houston, Texas

INTRODUCTION

The precursors for DNA synthesis, deoxyribonucleotides, are produced by direct reduction of the corresponding ribonucleotides in all organisms thus far examined. This fact is taken as supporting evidence for the hypothesis that life first evolved using an RNA-encoded information base. With the exception of lactobacilli, the reaction is of the form ribonucleoside diphosphate + reductant-$(SH)_2$ → deoxyribonucleoside diphosphate + reductant-(S-S), and is catalyzed by the enzyme ribonucleoside diphosphate reductase (ribonucleotide reductase).

Due to its central role in the control of DNA synthesis, ribonucleotide reductase has been extensively studied in a number of organisms. Among eukaryotes, the mammalian enzyme has been characterized in the greatest detail. It is composed of two nonidentical subunits, M1 and M2. The larger subunit, M1, has been purified to homogeneity from calf thymus and is a dimer of monomeric molecular weight 85,000 (Thelander and Berg, 1986; Thelander et al., 1985). Each monomer contains two distinct binding sites for deoxynucleoside triphosphates, which act as allosteric regulators of the enzymatic activity. One site controls the substrate specificity of the enzyme and is responsible for balancing the nucleotide pools. The other site measures the ATP/dATP ratio and controls the overall activity of the enzyme, presumably by ensuring that sufficient dNTPs are produced for DNA synthesis without completely depleting the ribonucleotides needed for RNA synthesis (Thelander et al., 1980). The smaller subunit, M2, is a dimer of monomeric molecular weight 44,000. Each monomer contains stoichiometric amounts of a non-heme iron center and a tyrosyl free radical which are essential for activity. These features are typical of other eukaryotic and viral ribonucleotide

We thank M. Kuroda and W. Harper for critical comments on the manuscript. We would also like to thank H. Hurd and J. Roberts for sharing unpublished data. This work was supported by grants NIGMS-1R01GM44664-01 and Q1186 from the Robert A. Welch Foundation to S. J. Elledge. Z. Zhou was a Robert A. Welch Predoctoral Fellow.

reductases and in this regard are similar to those of the enzymes from *Escherichia coli* and phage T4.

The genes encoding both subunits of ribonucleotide reductase from yeast have been cloned. The gene encoding the small catalytic subunit, *RNR2*, has been independently isolated by Elledge and Davis (1987) and Hurd et al. (1987). *RNR2* encodes a protein of 399 amino acids with an estimated molecular weight of 46,000. The protein shares 60% amino acid identity with the mammalian homologue (mouse), *RNR2* is essential for mitotic viability as determined by genetic disruption experiments using both tetrad analysis and a plasmid sectoring assay. Mutations that reduce *RNR2* function to a nonlethal level show hypersensitivity to hydroxyurea and to killing by methyl methanesulfonate (MMS) (Elledge and Davis, 1987). This latter fact demonstrates a role for ribonucleotide reductase in DNA repair processes.

The gene encoding the large subunit of the enzyme was cloned by using the mammalian gene as a probe and screening a yeast genomic library at reduced stringency (Elledge and Davis, 1990). Two distinct genes were isolated, and sequence analysis revealed that two homologous genes encode the yeast large subunit of ribonucleotide reductase. These genes, *RNR1* and *RNR3*, share 80% amino acid identity to each other and approximately 60% identity to the mammalian gene and map to different chromosomes. Northern analysis of RNA extracted from logarithmically growing cells showed that the *RNR1* gene encodes a 3.4-kb mRNA, but the *RNR3* mRNA could not be detected. Disruptions of the *RNR1* and *RNR3* genes in diploids followed by tetrad analysis revealed that *RNR1* is essential for mitotic viability. Spores bearing an *rnr1* null mutation show a *CDC* terminal phenotype upon germination of large budded cells, characteristic of mutations in genes required for DNA replication. Strains bearing an *RNR3* null mutation showed no phenotype. However, *RNR3* does encode a functional large subunit gene, because the *RNR3* gene when placed on high copy number 2-μm vector can suppress null mutations in the *RNR1* gene (Elledge and Davis, 1990). A summary of the *RNR* genes and their regulatory properties is shown in Table 1.

CELL CYCLE REGULATION

The eukaryotic cell cycle is a cascade of highly complex, sequential, independent, and interdependent events that culminate in the duplication of a cell. Many complex macromolecular structures must be assembled and disassembled with striking temporal and spatial precision. This degree of complexity necessitates the existence of a sophisticated regulatory network capable not only of coordinating these events, but also of correcting mistakes that occur during these complex processes. One level of this organization is accomplished through the restriction of cer-

TABLE 1. Summary of the *RNR* Genes Encoding Ribonucleotide Reductase

Gene	Subunit	mRNA size	Essentiality	DNA-damage inducibility	Cell-cycle regulation
RNR1	Large	3.4 kb	Yes	3 to 5-fold	10-fold
RNR2	Small	1.4 kb	Yes	25-fold	2-fold
RNR3	Large	3.4 kb	No	⩾100-fold	?

tain cell-cycle functions to particular periods of the cell cycle when they are needed. For example, DNA synthesis is restricted to a defined period of the cell cycle, S phase (Pringle and Hartwell, 1981). This restriction is accomplished, at least in part, through the temporal modulation of the activity and expression of gene products needed specifically in S phase. In *Saccharomyces cerevisiae* not only are genes encoding the enzymatic machinery for DNA synthesis cell-cycle-regulated [*POL1*, DNA polymerase 1 (Johnston et al., 1987), *CDC9*, DNA ligase (Peterson et al., 1985; Barker et al., 1985)], but so are several of the enzymatic activities involved in the production of the dNTP precursors needed for DNA synthesis [*CDC8*, thymidylate kinase (White et al., 1988), *CDC21*, thymidylate synthase (Storms et al., 1984)]. Ribonucleotide reduction was investigated by Lowden and Vitols (1973), who observed an activity maximum in S phase, and this activity was inhibited by hydroxyurea. The activity of the mammalian enzyme is also cell-cycle-regulated, being maximally expressed in S phase.

Several other genes associated with DNA metabolism are also cell-cycle-regulated [histones (Hereford et al., 1981), *HO*, an endonuclease involved in mating-type switching, *SWI5*, a regulator of mating-type switching (Nasmyth et al., 1987), and *RAD6* (Kupiec and Simchen, 1986)]. All of these genes are expressed in late G1 and/or S phase. In each case, with the exception of *RAD6*, which has not been tested, these genes are not expressed when cells are arrested in early G1 by a *cdc28* mutation or α-factor. The α-factor is a mating phermone that is capable of arresting cells at the start of the cell cycle in G1 phase. *CDC28* is a gene encoding a protein kinase that controls the G1–S phase transition in *S. cerevisiae*. Temperature-sensitive mutations in the *CDC28* gene cause cells to arrest the cell cycle in G1 at a point indistinguishable from the α-factor arrest point. *HO*, *CDC8*, *CDC9*, *CDC21*, and *POL1* are expressed immediately upon reaching the start of the cell cycle in late G1, before the *CDC4* block. *CDC4* is temporally the earliest function that is known to arrest the cell cycle after start but before S phase. The *HTA1* and *HTB1* genes encoding histones H2A and H2B are expressed in S phase after the *CDC4* block. This regulation could occur at the level of mRNA synthesis or degradation, or both. In yeast, the only genes that have been examined so far, *HO* (Nasmyth, 1985), *CDC21* (McIntosh et al., 1988), and *HTA1* and

HTB1 (Osley et al., 1986), are regulated at the level of mRNA synthesis, although posttranscriptional regulation is important for histone regulation in higher eukaryotic cells (Schumperli, 1986). Positive- and negative-acting sequences are involved in cell-cycle regulation of the histone *HTA1* and *HTB1* genes (Osley et al., 1986), and mutations in the *HIR* genes have been identified that alter the negative regulation (Osley and Lycans, 1987). The *hir* mutations do not affect the regulation of the late G1 class of genes. The *HO* gene encodes an endonuclease that is involved in mating type switching. It is under several other types of control in addition to cell-cycle regulation and has an extremely complicated regulatory region. In the *HO* promoter a repeated sequence, $CACGA_4$, has been found to confer cell-cycle regulation on a heterologous promoter (Nasmyth, 1984). This repeat has not been found in any other cell-cycle-regulated genes. Two genes, *SWI4* and *SWI6*, have been identified that disrupt the cell-cycle regulation of *HO*. When either gene is mutant, *HO* is not expressed and cells fail to switch their mating type. A DNA-binding factor has been identified that binds this sequence in an *SWI4-* and *SWI6-* dependent fashion (Andrews and Herskowitz, 1989), but if and how that factor is cell-cycle regulated is not known.

Cell-cycle regulation of the *RNR* genes has been investigated by Elledge and Davis (1990). Northern analysis using message prepared from populations of cells moving synchronously through the cell cycle revealed that the transcript for *RNR1* fluctuated greater than 10-fold in the cell cycle, while the *RNR2* transcript showed a modest twofold fluctuation relative to *URA3* mRNA, which is constant through the cell cycle. Therefore, the cell cycle fluctuation of ribonucleotide reductase activity is likely to be primarily due to modulation of the *RNR1* transcript, although studies of the protein levels must be performed to strictly prove this point. Transcription of *RNR1* is coordinately regulated with transcription of the *POL1* gene that encodes DNA polymerase 1. Both are induced in late G1 phase, directly preceding the initiation of DNA replication. This makes a certain amount of biological sense, because both enzymes are required at precisely the same time in the cell cycle, at the start of S phase.

In a further analysis of the cell-cycle regulation of *RNR1* and *POL1*, Elledge and Davis (1990) found that the ability to induce the *RNR1* and *POL1* transcripts after arrest in G1 is dependent on protein synthesis that can be blocked by cycloheximide. This may explain the early observations of Hereford and Hartwell (1974) that progression from the α-factor-induced G1 arrest to S phase required protein synthesis. Although further analysis is required to determine whether this is a general requirement of all G1 arrested states such as the *cdc28* arrest point, or is specific to the α-factor arrest, it does raise some interesting possibilities. One possibility is that a critical level of a protein product must accumulate for progression into S and activation of transcription of S phase specific genes, and that the mechanism through which α-factor arrests the cell cycle is by specifically depleting or altering this protein. Cycloheximide would then act to block the accumulation of that protein after release from the α-factor block. The best candidate for such a protein is cyclin. Cyclins were identified because of their unusual behavior during the cell cycle (Evans et al., 1983). Cyclins accumulate during the cell cycle, peaking in abundance at the end of G2. After cells enter mitosis, the cyclins are rapidly destroyed. Cyclins are thought to be involved in the activation of the p34 protein kinase. This protein is conserved through evolution and is homologous to the CDC28 protein kinase of S. cerevisiae. CDC28 regulates entry into S phase in S. cerevisiae, and three cyclin homologs have been identified that appear to be involved in this regulatory step, CLN3, also known as DAF1 and WHI1 (Cross, 1988; Nash et al., 1988), CLN1, and CLN2 (Hadwiger et al., 1989). CLN3 has been implicated in the α-factor arrest pathway (Courchesne et al., 1989; Cross, 1988). Depletion or physical alteration of cyclins could occur during α-factor arrest, thereby blocking the activation of CDC28, causing cell-cycle arrest. Recovery would then be mediated by the cycloheximide-sensitive synthesis of new, unaltered cyclins.

The fact that the late G1 class of genes is expressed immediately after, and is dependent on, the start of the cell cycle suggests that the trans-acting factors responsible for their expression may be sensing this early control event directly. The other genes such as the histones are expressed later in the cell cycle and therefore likely to be sensing a different signal. These cell-cycle-regulated genes now serve as molecular markers for cell-cycle regulatory events, in addition to the obvious landmark cytological events such as bud emergence, nuclear migration, spindle formation, and cytokinesis. Since the methods for analyzing transcriptional regulation are highly developed, these genes provide an excellent opportunity for use as tools to probe the infrastructure of the cell-cycle regulatory circuitry. In particular, identification of the earliest signals in the G1 to S phase transition could be of substantial significance, revealing potential interfaces between oncogene-controlled regulatory pathways and the cell cycles of higher eukaryotes.

Sequences That Mediate the Cell-Cycle Control of *RNR1*

Sequence comparison between the *RNR1* and *RNR3* promoter regions identified a duodecamer repeat of sequence A/T A/T A/T AGCGCT A/T A/T A/T that is repeated twice in *RNR3* and four times in the *RNR1* promoter (Elledge, unpublished observation). These repeats contain at their core the recognition site for the restriction enzyme *Mlu* I. A computer search of the yeast data base for genes that contain at least two repeats of the core consensus AGCGCT yielded only two genes out of over 500 kb of sequence: *CDC21* (thymidylate synthase) and *POL1* (DNA polymerase 1). Both of these genes are S-phase specific and coordinately regulated with the *RNR1* transcript. In addition, two other S-phase-specific genes, *CDC8* (thymidylate kinase) and *CDC9* (DNA ligase), contain one copy of the repeat, and a new unpublished gene *PRI2*, encoding a subunit of the primase for DNA polymerase, is reported to have two such repeats and to be S-phase specific (Dr. Plevliani, personal communication). The positions of these elements in the promoters of S-phase-specific genes are illustrated in Fig. 1. A deletion of 50 bp in the *CDC21* promoter that destroys these two repeats abolishes cell-cycle regulation (McIntosh et al., 1988). Oligonucleotides representing the core of the consensus sequence were synthesized and shown to act as an upstream activating sequence (UAS) when present in multiple copies in front of the *CYC1* promoter (Elledge, unpublished observation; McIntosh et al., 1991; Lowndes et al., 1991). The strength of transcription increased with increasing copy number of the oligonucleotide inserts. Evidence from the studies of the *CDC21* promoter suggests that these sequences are controlling the cell-cycle regulation (McIntosh et al., 1991). A DNA binding protein that binds these sequences has been detected in crude extracts, and the binding pattern shows a cell-cycle periodicity (Lowndes et al., 1991) that may be responsible for the cell-cycle regulation, although this remains to be proven. It should be noted that it is not yet known

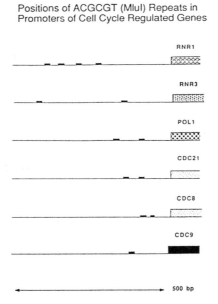

Positions of ACGCGT (MluI) Repeats in
Promoters of Cell Cycle Regulated Genes

FIGURE 1. Positions of the *Mlu*I cell-cycle regulation motif in the promoters of S-phase-specific genes. The small black box indicates the position of core consensus elements in the promoters of the indicated genes. For *CDC8*, the second box closest to the structural gene is not a complete match, differing in the last base of the consensus sequence.

Regulation of S-Phase Specific Genes

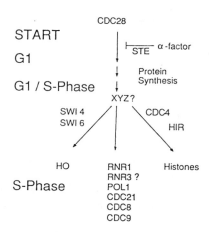

FIGURE 2. Summary of the biochemical and genetic events between start and the induction of S-phase-specific gene expression. The names of genes involved are in capital letters. Only those genes known to have a regulatory role are listed. Other genes that operate at start are presumed to have a role but are not listed because their role in the induction of S-phase genes has not yet been tested.

whether the transcription from the *RNR3* gene is cell-cycle regulated when it is expressed, although it has two copies of the repeat sequence. A summary of the pathway for induction of S-phase gene expression is shown in Fig. 2.

DNA-Damage Inducible Genes

Multiple forms of regulation are often found in the cell-cycle-regulated genes because, although these genes are required primarily in one period of the cell cycle, circumstances can arise in which their expression is needed outside of the period in which they are normally expressed. For example, repair of certain types of DNA damage requires the ability to synthesize DNA. If repair is to proceed outside of S phase, then the cell must possess the capacity to synthesize DNA in other phases of the cell cycle. In fact, several genes with cell-cycle-regulated activities involved in DNA synthesis are inducible by DNA damage [*POL1* (Johnston et al., 1987), *CDC9* (Barker et al., 1985), *CDC8* (Elledge and Davis, 1987), and all three genes encoding ribonucleotide reductase, *RNR1*, *RNR2*, and *RNR3* (Elledge and Davis, 1987, 1989a, 1989b, 1990; Hurd et al., 1987)]. The induction of these genes in response to the stress of DNA damage is thought to produce a metabolic state that facilitates DNA replicational repair processes. The overall objective of the cell-cycle regulation and the DNA-damage regulation is the same, to provide the capacity to synthesize DNA when it is needed. Therefore, these different regulatory pathways may share common components. Perhaps the DNA-damage sensory network activates gene expression via the existing cell-cycle regulatory circuit. This is an important possibility and must be explored.

What is known about the DNA-damage response? In *E. coli*, treatment with agents that damage DNA or block replication causes the appearance of a set of physiological responses including the induction of DNA repair processes, mutagenesis, and induction of lysogenic bacteriophage (reviewed in Walker, 1985). These processes have collectively been called the SOS response because at least some of them appear to promote cell survival. In all, over 20 genes have been identified that are activated transcriptionally in response to DNA damage. The molecular mechanism of this coordinately regulated response involves the proteolytic inactivation of a common repressor, the LexA protein, by an activated form of the RecA protein. Much less is known about the response to DNA damage in eukaryotes. In *S. cerevisiae* a number of genes have been identified on the basis of increased transcription in response to DNA damage, the *DIN* (Ruby and Szostak, 1985) and *DDR* (McClanahan and McEntee, 1984) genes. However, the function of these genes remains unknown. Several genes of known function have also been demonstrated to show DNA-damage inducibility, including *RAD2* (excision repair) (Robinson et al., 1986), *RAD54* (recombinational repair) (Cole et al., 1987), and *UBI4* (protein degradation) (Treger et al., 1988), in addition to the cell-cycle-regulated genes already mentioned. It has not been determined whether these genes are regulated by the same DNA-damage-sensing pathways. In addition, cells exhibit a *RAD9*-dependent response to DNA damage that arrests the cell cycle in G2. Cells mutant in the *RAD9* gene fail to arrest their cell cycle in response to X-rays. Although the *RAD9* transcript is not damage-inducible, the *RAD9* gene must sense DNA damage.

Little is known about the mechanisms that sense and respond to DNA damage in eukaryotes. Although several inducible genes have been identified, it has yet to be demonstrated that the induction of any gene is important physiologically in the cellular response to damage. Furthermore, no mutants are available that block the inducibility of these genes.

DNA-Damage Inducibility of the *RNR* Genes

RNR1, *RNR2*, and *RNR3* are all inducible at the level of transcript accumulation by agents that block DNA synthesis, such as hydroxyurea (HU) and methotrexate, or by agents that damage DNA, such as UV light, MMS, and 4-nitroquinoline 1-oxide (4-NQO). *RNR1* is inducible three- to fivefold, *RNR2* is inducible 25-fold, and *RNR3* is inducible 100- to 500-fold. *RNR3* was found to be identical to the previously isolated gene *DIN1* (Yagle and McEntee, 1990), which was isolated by its ability to be induced by DNA damaging agents. Under normal vegetative conditions, only *RNR1* and *RNR2* are expressed, and thus the ribonucleotide reductase in the cell is of a homogeneous form,$\alpha 1_2 \beta_2$. However, in the presence of DNA damage, the second large subunit gene is induced, producing at least one, $\alpha_2 \beta_2$, and possibly two, $\alpha 1 \alpha 2 \beta_2$, additional forms of ribonucleotide reductase differing in their subunit composition. The physiological role of *RNR3* remains to be determined. The yeast strains mutant for *RNR3* have no obvious growth defects, and are not sensitive to HU or DNA-damaging agents (Elledge and Davis, unpublished). Furthermore, the yeast strain that lacks *RNR1* and overproduces *RNR3* is not sensitive to HU (Elledge, unpublished), suggesting that the *RNR3* protein can substitute efficiently for the *RNR1* protein when overproduced. What then is the role of *RNR3*? Clearly, it must confer some selective advantage for the cell to have conserved both its function and tight regulation. One possibility is that the *RNR3* protein has altered regulatory properties allowing the cell to survive certain types of stress that we have yet to duplicate in the laboratory. For example, biological fungicides are rampant in the wild, and are often targeted to inhibit key regulatory enzymes. Perhaps the evolutionary history of *S. cerevisiae* included growth in an eco-system in which inhibitors of ribonucleotide reductase were a commonly employed competitive strategy. Duplication of the target gene would facilitate the rapid evolution of drug resistant variants. Alternatively, the *RNR3* gene may play a role in a nonvegetative function of yeast such as meiosis. Recent studies using a new reporter gene system capable of recording the history of gene expression in a colony have revealed that *RNR3* expression is induced in late stages of colony growth and perhaps is responding to anaerobic conditions on the interior of colonies (Elledge, manuscript in preparation). Thus, *RNR3* may have a role under anaerobic conditions or in stationary-phase existence. *RNR3* may play a marginal role in cell survival under certain conditions that are not immediately obvious in short-term laboratory experiments but that stand a greater chance of detection in long-term chemostatic types of experiments.

The DNA-damage inducibility of the *RNR2* gene has been studied extensively. This analysis has revealed that induction of *RNR2* by DNA damage is independent of protein synthesis that is inhibited by cycloheximide and is not dependent on cell cycle stage, and that sequences from the regulatory region are able to confer DNA-damage inducibility upon heterologous promoters (Elledge and Davis, 1989b; Hurd and Roberts, 1989). Insensitivity to inhibition by cycloheximide formally eliminates the de novo synthesis of a positive activator as a model for DNA damage induction, and leaves two equally plausible molecular models, the post-translational modification of a positive activator or the inactivation of a negatively acting factor.

What are the signals to which the *RNR2* promoter responds? Clearly

nucleotide depletion can induce *RNR2*, because treatment with methotrexate or hydroxyurea can induce high levels of synthesis of *RNR2*
mRNA. It makes a certain amount of biological sense that an enzyme
whose role is to synthesize deoxyribonucleotides would be induced in
response to their depletion. It is also possible that the DNA damage
induction proceeds through nucleotide depletion brought about by repair synthesis. To test this hypothesis, Elledge and Davis (1989a) examined the inducibility of *RNR2* by 4-nitroquinoline-1-oxide (4-NQO) in a
wild type and *rad4*-2 mutant background. 4-NQO is a UV-mimetic agent
that produces bulky DNA adducts that are repaired through the excision
repair pathway. The *rad4*-2 mutant blocks the incision step of excision
repair and thus totally blocks excision repair and its capacity to deplete
deoxyribonucleotide pools. In this experiment *RNR2* induction was not
blocked by the *rad4* mutation. In fact, the *rad4* mutation made *RNR2*
hyperinducible to low levels of 4-NQO, implying that the regulators of
RNR2 are actually responding to the levels of DNA damage present.
Whether nucleotide depletion and DNA damage signals are identical and
are mediated via the same proteins remains to be determined.

Detailed deletion analysis of the *RNR2* regulatory region has implicated both a positive and a negative regulatory element in the response
and identified a 70-bp fragment called the DRE (DNA-damage-responsive
element) that can confer some DNA-damage inducibility upon a heterologous promoter. A schematic drawing of the positions of DNA-binding
proteins on this element is shown in Fig. 3. Originally there were thought
to be four proteins that bound this element, RRF1, 2, 3, and RAP1 (GRF1).
RRF refers to ribonucleotide reductase regulatory factors. RAP1 (repressor and activator protein) (Buchman et al., 1988a, 1988b; Shore and Nasmyth, 1987) is an abundant multifunctional protein that can act to activate gene expression, as is the case in many yeast promoters including
the *MAT* locus and many of the ribosomal gene promoters. It can also act
as a negative factor in particular contexts to repress transcription, as in
the case of the silent mating-type loci (*HMR a* and *HML α*). However,
recent experiments by Zhou and Elledge (unpublished) have shown that
the RRF1 was RAP1, and there are now two RAP1 binding sites in this
promoter. Alteration of the pattern of DNA binding proteins, their abundance or affinity does not appear to occur in the presence of DNA damage. Hurd and Roberts (1989) clearly demonstrated that the second RAP1
site at −356 (Fig. 3) is not essential for the response to DNA damage. It is
clear that removal of the second site, which is a much higher affinity
binding site than site 1, reduces the basal level of transcription, and it is
thought to act in a positive fashion, although it is nonessential for *RNR2*
induction by DNA-damaging agents (Hurd and Roberts, 1989). The role of
the upstream site (−403, Fig. 3) has not been conclusively addressed at
present. It is near or coincident with a repressing sequence that can act
to repress transcription. This repressing sequence is located on a 42-bp

DNA Damage Responsive Element of RNR2

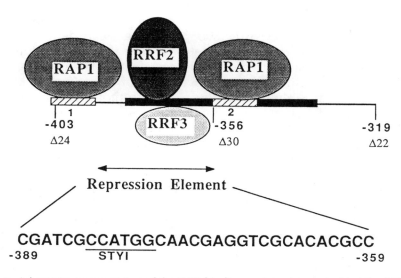

FIGURE 3. Schematic representation of the DNA binding proteins known to bind the DRE (DNA-damage-responsive element) of the *RNR2* promoter. The precise locations of the RAP1 binding sites are known. The precise locations of the binding sites for RRF2 and RRF3 (*RNR* regulatory factors) are not known within the 47-bp element between −403 and −356. RRF3 is thought to bind more closely to the −356 sequence, although footprinting studies are needed to establish this location (Elledge and Davis, 1989b). The numbering system employed here is that of Hurd and Roberts (1989) and designates the start of transcription as +1. The sequence shown in the lower section is the smallest region that contains the repressing element of *RNR2* as defined by Hurd and Roberts (1989) and Elledge and Davis (1989b).

fragment of DNA and can act to confer negative regulation on heterologous activating sequences such as the *MAT* α UAS (Elledge and Davis, 1989b). This negatively acting sequence can repress when positioned either 5′ to a UAS or between a UAS and the TATA sequence, although in the latter position it is much more effective, repressing up to 100-fold. The *MAT* α UAS is a RAP1 binding site. Hurd and Roberts (personal communication) have narrowed down the repression sequence to 30 bp (−389 to −359 bp) and have shown that it can act to repress the *CYC1* UAS region, which does not contain a RAP1 binding site. Furthermore, they have observed cooperativity in repression, with one element yielding 15-fold repression, while two tandem elements produce greater than 1000-fold repression. Neither group has been able to detect DNA-damage induction of these repressed heterologous promoters (Elledge and Davis, 1989b; Hurd and Roberts, personal communication). Both groups have determined that a *Sty*I site in the sequence is essential for repression, but a protein that binds to this site has not been detected (Elledge and Davis,

1989b). Thus the precise role of the negative element in *RNR2* regulation remains to be elucidated.

Detailed analysis of the deletion studies on *RNR2* suggests that more than one element may be capable of sensing DNA damage. The DRE element of Elledge and Davis (1989b) can confer DNA damage inducibility upon a *lacZ* reporter gene. Deletions from the 5' end that remove this sequence (and all upstream sequences) destroy DNA-damage inducibility. However, deletions that remove most of this element but retain sequences 5' to this element in the native promoter retain some DNA-damage inducibility, suggesting that there are sequences upstream of the DRE that are capable of sensing DNA damage. This issue may be further complicated by the fact that *RNR2* shows weak cell-cycle regulation and that most DNA-damaging treatments elongate S phase, giving rise to artificial synchronization.

The *RNR1* and *RNR3* promoters have not been subjected to as extensive deletion analysis as the *RNR2* promoter. Unpublished analysis of the *RNR1* promoter has identified several positively acting sequences and one negative sequence element. When the entire upstream region of *RNR1* is placed upstream of a *CYC1-lacZ* reporter construct, the three- to fivefold DNA-damage-inducibility response can be observed. Furthermore, by subcloning smaller fragments of this promoter, a 100-bp fragment has been identified that has a lower basal level but that can confer 25-fold DNA-damage inducibility upon the *CYC1-lacZ* reporter gene (Elledge, unpublished). This is interesting because it suggests that elements in the *RNR1* promoter can sense DNA damage and alter their activity over a large range; however, only three- to fivefold induction can be observed due to the high basal level of the native promoter. Thus, small induction ratios in genes responding to any type of stimulus may not truly reflect the regulatory capacity of these elements. *RNR1* may be a good example of the complex interplay of multiple types of regulation, both cell-cycle and DNA-damage regulation. The 100-bp fragment that confers the 25-fold DNA damage inducibility has many of the *Mlu*I cell-cycle motifs removed, suggesting that they may be independent regulatory elements. Preliminary deletion analysis of the *RNR3* promoter has revealed both positive and negative regulatory elements (Zhou and Elledge, unpublished).

FUTURE DIRECTIONS

The direction of future analysis of the cell-cycle regulation of the *RNR* gene family will likely follow the analysis of protein factors that interact with the *Mlu*I sequence elements. Isolation and analysis of the gene(s) encoding this binding protein(s) will be a critical step forward in the analysis of cell-cycle-regulated transcription of S-phase genes. In addition, the determination of how these factors interface with the cell-cycle regulatory machinery is likely to proceed rapidly once the appropriate genes are identified. In particular, the role of the *CLN* genes and *CDC28* in the activation of S-phase gene expression must be explored at a molecular and genetic level. Are the activators of S-phase gene expression direct substrates for the *CDC28* protein kinase? Are other gene products also required at start, such as *CDC36* and *CDC39*, in addition to these proteins? These questions await molecular and genetic elucidation.

It is clear that there remain many unanswered questions with respect to the regulation of the *RNR* genes by DNA damage. The detailed analysis of the *RNR2* promoter has generated more questions than answers. The questions of multiple DNA-damage-sensing elements, of modes of regulation (positive or negative), and of the identity of proteins mediating this response all remain unresolved issues requiring more detailed analysis. What is clear is that the regulatory region of *RNR2* is much more complicated than anticipated, containing multiple binding sites for proteins of as yet undetermined function. The role of these proteins may be revealed by site-directed mutagenesis studies of the DRE element. The function of the negatively acting sequence must also be explored. Although it is clear that it cannot mediate damage inducibility alone, it may still play a role. Site-directed mutagenesis will be required to unravel the role of the various binding factors of the DRE. Furthermore, it is possible that because so many cell-cycle-regulated genes are DNA-damage inducible, there is some overlap with the regulatory pathways. This possible link must be explored.

In a larger sense, the transcription factors that mediate the DNA damage response, although important, are not likely to be the entire answer to the regulation of these genes. So far, of the binding factors identified, none have been shown to bind two of the *RNR* promoters (Elledge, unpublished), suggesting either that the single factor that mediates the DNA damage signal has not yet been detected biochemically, or that multiple nonoverlapping factors are capable of transducing this signal to each promoter. To identify the factors that sense and transduce this DNA damage response it will be necessary to go beyond the transcription factors to identify those molecules that sense and transduce the stress signal produced by DNA damage. Thus, the future of this endeavor lies in the identification of trans-acting mutations that alter the regulation of these genes in response to DNA damage. Toward this end, Zhou and Elledge (1992) have isolated spontaneous trans-acting mutants that express *RNR3* constitutively, the *crt* (constitutive *RNR* transcription) mutants. Two hundred mutants have been isolated, of which 7% are also temperature-sensitive lethal at 37°C. These mutations fall into 9 complementation groups in an ongoing analysis. It is expected that many of the constitutive mutations will lie in genes that only indirectly induce *RNR3* by creating a situation in which a stress signal is generated. For example, mutations that reduce the function of *RNR2* cause an increase in expression of *RNR2-lacZ* reporter constructs (Elledge and Davis, 1989a). In this case it is unlikely that *RNR2* itself is directly involved in mediating the signal to increase its expression and that the resulting increase in expression is due to an indirect effect of lower nucleotide levels. Many mutants of this type are expected in the constitutive group, along with those directly involved in the signal transduction itself. The challenge of future work is to distinguish between these constitutive classes of mutations and to identify additional mutations that fail to induce *RNR3* in response to DNA damage. Uninducible mutations are likely to be more directly involved in the signal transduction pathway.

The capacity of cells to efficiently sense and respond to environmental stress is central to an organism's ability to survive and evolve. The analysis of the regulation of DNA-damage inducible genes will shed light on how these sensory mechanisms operate and may reveal further connections to the cell-cycle-regulatory machinery. If this sensory mechanism is shared among eukaryotes, as is likely to be the case, this information may have a significant impact on our understanding of how human cells sense and respond to the presence of DNA-damage agents.

REFERENCES

Andrews, B. J., and Herskowitz, I. 1989. Identification of a DNA-binding factor involved in cell cycle control of the yeast HO gene. *Cell* 57:21–29.

Barker, D. G., White, J. M., and Johnston, L. H. 1985. The nucleotide sequence of the DNA ligase gene (CDC 9) from *Saccharomyces cerevisiae*: A gene which is cell-cycle regulated and induced in response to DNA damage. *Nucleic Acids Res.* 13:8323–8337.

Buchman, A. W., Kimmerly, W. J., Rine, J., and Kornberg, R. D. 1988a. Two DNA-binding factors recognize specific sequences at silencers, upstream activating sequences, autonomously replicating sequences, and telomeres in *Saccharomyces cerevisiae*. *Mol. Cell. Biol.* 8:210–225.

Buchman, A. W., Lue, N., and Kornberg, R. D. 1988b. Connections between transcriptional activators, silencers, and telomeres as revealed by functional analysis of a yeast DNA-binding protein. *Mol. Cell. Biol.* 8:5086–5099.

Cole, G. M., Schild, D., Lovett, S. T., and Mortimer, R. K. 1987. Regulation of RAD54- and RAD52-lacZ gene fusions in *Saccharomyces cerevisiae* in response to DNA damage. *Mol. Cell. Biol.* 7:1078–1084.

Courchesne, W. E., Kunisawa, R., and Thorner, J. 1989. A putative protein kinase overcomes pheromone-induced arrest of cell cycling in *S. cerevisiae*. *Cell* 58:1107–1119.

Cross, F. R. 1988. DAF1, a mutant gene affecting size control, phermone arrest, and cell cycle kinetics of *Saccharomyces cerevisiae*. *Mol. Cell. Biol.* 8:4675–4684.

Elledge, S. J., and Davis, R. W. 1987. Identification and isolation of the gene encoding the small subunit of ribonucleotide reductase from *Saccharomyces cerevisiae*: A DNA damage-inducible gene required for mitotic viability. *Mol. Cell. Biol.* 7:2783–2793.

Elledge, S. J., and Davis, R. W. 1989a. DNA-damage induction of ribonucleotide reductase. *Mol. Cell. Biol.* 9:4932–4940.

Elledge, S. J., and Davis, R. W. 1989b. Identification of a damage regulatory element of RNR2, and evidence that four distinct proteins bind to it. *Mol. Cell. Biol.* 9:5373–5386.

Elledge, S. J., and Davis, R. W. 1990. Two genes, differentially regulated by DNA damage and the cell cycle, encode alternate regulatory subunits of ribonucleotide reductase. *Genes Dev.* 4:740–751.

Evans, T., Rosenthal, E. T., Youngblom, J., Distel, D., and Hunt, T. 1983. Cyclin: A protein specified by maternal mRNA in sea urchin eggs that is destroyed at each cleavage division. *Cell* 33:389–396.

Hadwiger, J. A., Wittenberg, C., and Mendenhall, M. D. 1989. A novel family of cyclin homologs that control G1 in yeast. *Proc. Natl. Acad. Sci. USA* 86:6255–6259.

Hereford, L. M., and Hartwell, L. H. 1974. Sequential gene function in the initiation of *Saccharomyces cerevisiae* DNA synthesis. *J. Mol. Biol.* 84:445–461.

Hereford, L. M., Osley, M. A., Ludwig, J. R., and McGlaughlin, C. S. 1981. Cell-cycle regulation of yeast histone mRNA. *Cell* 24:367–375.

Hurd, H., and Roberts, J. 1989. Upstream regulatory sequences of the yeast RNR2 gene include a repression sequence and an activation site that binds the RAP1 protein. *Mol. Cell. Biol.* 9:5366–5372.

Hurd, H. K., Roberts, C. W., and Roberts, J. W. 1987. Identification of the gene for the yeast ribonucleotide reductase small subunit and its inducibility by methyl methanesulfonate. *Mol. Cell. Biol.* 7:3673–3677.

Johnston, L. H., White, J. H. M., Johnson, A. L., Lucchini, G., and Plevani, P. 1987. The yeast DNA polymerase I transcript is regulated in both the mitotic cell cycle and in meiosis and is also induced after DNA damage. *Nucleic Acids Res.* 15:5017–5030.

Kupiec, M., and Simchen, G. 1986. Meiotic and mitotic regulation of RAD6. *Mol. Gen. Genet.* 203:538–543.

Lowden, M., and Vitols, E. 1973. Ribonucleotide reductase activity during the cell cycle of *Saccharomyces cerevisiae.* *Arch. Biochem. Biophys.* 158:177–184.

Lowndes, N. F., Johnson, A. L., and Johnston, L. H. 1991. Coordination of DNA synthesis genes in budding yeast by a cell-cycle regulates *trans* factor. *Nature* 350:247–250.

McClanahan, T., and McEntee, K. 1984. Specific transcripts are elevated in *Saccharomyces cerevisiae* in response to DNA damage. *Mol. Cell. Biol.* 4:2356–2363.

McIntosh, E. M., Ord, R. W., and Storms, R. K. 1988. Transcriptional regulation of the cell cycle-dependent thymidylate synthase gene of *Saccharomyces cerevisiae.* *Mol. Cell. Biol.* 8:4616–4624.

McIntosh, E. M., Atkinson, T., Storms, R. K., and Smith, M. 1991. Characterization of a short, *cis*-acting DNA sequence which conveys cell cycle stage-dependent transcription in *Saccharomyces cerevisiae.* *Mol. Cell. Biol.* 11:329–337.

Nash, R., Tokiwa, G., Anand, S., Erikson, K., and Futcher, B. 1988. The WHI1+ gene of *Saccharomyces cerevisiae* tethers cell division to cell size and is a cyclin homolog. *EMBO J.* 7:4335–4346.

Nasmyth, K. 1985. A repetitive DNA sequence that confers cell-cycle START (CDC28)-dependent transcription of the HO gene. *Cell* 42:225–235.

Nasmyth, K., Seddon, A., and Ammerer, G. 1987. Cell cycle regulation of SWI5 is required for mother-cell-specific HO transcription in yeast. *Cell* 49:549–558.

Osley, M. A., Gould, J., Kim, S., Kane, M., and Hereford, L. 1986. Identification of sequences in a yeast histone promoter involved in periodic transcription. *Cell* 45:537–544.

Osley, M. A., and Lycans, D. 1987. Trans-acting regulatory mutations that alter transcription of *Saccharomyces cerevisiae* histone genes. *Mol. Cell. Biol.* 7:4204–4210.

Peterson, T. A., Prakash, L., Prakash, S., Osley, M. A., and Reed, S. I. 1985. Regulation of CDC9, the *Saccharomyces cerevisiae* gene that encodes DNA ligase. *Mol. Cell. Biol.* 5:226–235.

Pringle, J. R., and Hartwell, L. H. 1981. The *Saccharomyces cerevisiae* cell cycle. In *The Molecular Biology of the Yeast Saccharomyces cerevisiae,* eds. J. N. Strathern, E. W. Jones, and J. R. Broach, pp. 97–142. Cold Spring Harbor, N.Y.: Cold Spring Harbor Laboratory.

Robinson, G., Nicolet, C., Kalainov, D., and Friedberg, E. 1986. A yeast excision-repair gene is inducible by DNA damaging agents. *Proc. Natl. Acad. Sci. USA* 83:1842–1846.

Ruby, S. W., and Szostak, J. W. 1985. Specific *Saccharomyces cerevisiae* genes are expressed in response to DNA-damaging agents. *Mol. Cell Biol.* 5:75–84.

Schumperli, D. 1986. Cell-cycle regulation of histone gene expression. *Cell* 45:471–472.

Shore, D., and Nasmyth, K. 1987. Purification and cloning of a DNA-binding protein that binds to both silencer and activator elements. *Cell* 51:721–732.

Storms, R. K., Ord, R. W., Greenwood, M. T., Mirdamadi, B., Chu, F. K., and Belfort, M. 1984. Cell-cycle dependent expression of thymidylate synthase in *Saccharomyces cerevisiae*. *Mol. Cell. Biol.* 4:2858–2864.

Thelander, L., and Berg, P. 1986. Isolation and characterization of expressible cDNA clones encoding the M1 and M2 subunits of mouse ribonucleotide reductase. *Mol. Cell. Biol.* 6:3433–3442.

Thelander, L., Erikson, L., and Akerman, M. 1980. Ribonucleotide reductase from calf thymus. *J. Biol. Chem.* 255:7426–7432.

Thelander, M., Graslund, A., and Thelander, L. 1985. Subunit M2 of mammalian ribonucleotide reductase. *J. Biol. Chem.* 260:2737–2741.

Treger, J. M., Heichman, K. A., and McEntee, K. 1988. Expression of the yeast UBI4 gene increases in response to DNA-damaging agents and in meiosis. *Mol. Cell. Biol.* 8:1132–1136.

Walker, G. C. 1985. Inducible DNA repair systems. *Annu. Rev. Biochem.* 54:425–457.

White, J., Green, S. R., Barker, D. G., Dumas, L. B., and Johnston, L. H. 1988. The CDC8 transcript is cell cycle regulated in yeast and is expressed coordinately with CDC9 and CDC21 at a point preceding histone transcription. *Exp. Cell Res.* 171:223–231.

Yagle, K., and McEntee, K. 1990. The DNA damage-inducible gene *DIN1* of *S. cerevisiae* encodes a regulatory subunit of ribonucleotide reductase and is identical to *RNR3*. *Mol. Cell. Biol.* 10:5553–5557.

Zhou, Z., and Elledge, S. J. 1992. Isolation of *crt* mutants constitutive for transcription of the DNA damage inducible gene *RNR3* in *Saccharomyces cerevisiae*. *Genetics* (in press).

6 | REVIEW: GENE AMPLIFICATION—A CELLULAR RESPONSE TO GENOTOXIC STRESS

Christine Lücke-Huhle

Kernforschungszentrum Karlsruhe, Institut für Genetik und Toxikologie, Karlsruhe, Germany

Although environmental agents play an important role in human carcinogenesis, little is known about the initial molecular events leading to their carcinogenic action. Tumor cells differ from normal cells by exhibiting characteristic genomic alterations. These include mutations, chromosomal aberrations and rearrangements, deletions, and gene amplifications. The increase in gene copy number of one or a few genes, called gene amplification, is a mechanism leading to increased expression of normal or altered gene products and is thought to participate in carcinogenesis (Zur Hausen and Schlehofer, 1987).

Genes are amplified in many organisms. Examples have been described for bacteria (Anderson and Roth, 1977), plants (Shah et al., 1986), insects (DeCicco and Spradling, 1984), and a variety of mammalian cell systems (reviews: Schimke, 1984a; Stark et al., 1989). DNA amplification occurs physiologically in normal cell regulation and differentiation (Long and Dawid, 1980; Spradling and Mahowald, 1980) and is also the result of genotoxic insult. In particular, various types of radiation (Lücke-Huhle et al., 1986; Ehrfeld et al., 1986; Lücke-Huhle and Herrlich, 1986) and chemical carcinogens (Lavi, 1982; Pool et al., 1989) efficiently induce gene amplification in cultured rodent cells.

Most of the mechanistic studies on gene amplification have been done on drug resistance. Here the relationship between amplification of a specific gene and overcoming growth constraint of the selecting agent by an increased production of the gene product is understood (Schimke, 1984b; Stark and Wahl, 1984). The physiological consequences of amplified cellular oncogenes, however, are not yet clarified. They are found in a variety of human tumors (Schwab and Amler, 1990) and are commonly believed to be the basis for unlimited growth in that they allow the host cell to escape growth control, become mobile and invasive, or escape immune surveillance.

Two unusual cytological phenomena have been associated with amplified DNA (Biedler and Spengler, 1976; Balaban-Malenbaum and Gilbert, 1977): first, the presence of small acentric chromosomes, which,

The author thanks Sabine Mai for helpful discussions and Professor Peter Herrlich for critical reading of the manuscript.

because they often appear in pairs, are called double minutes (DMs); and second, the appearance of long chromosome segments with little evidence of G-banding, referred to as homogeneously staining regions (HSRs).

Amplification could be caused by dysfunction of any one of the many steps during replication. Most likely, the decisive step is found in the initiation reaction of replication. The induction of oncogene amplification (Ehrfeld et al., 1986) is not sensitive enough for a study of initial events. However, amplification of viral sequences integrated in the mammalian genome can be traced in most of the treated cells at a very early time, possibly due to participation of viral proteins such as SV40 T antigen. We therefore used this system in order to characterize initiating events of gene amplification. The viral system resembles that of cellular amplification in that similar karyotypic characteristics such as DMs and variable lengths of amplified DNA in different amplification events are seen.

Our interest has been centered on three questions:

1. What are the initial events in the signal transfer pathway following DNA damage?
2. What are the genomic consequences of amplified DNA sequences?
3. Can diploid human cells amplify genes as easily as is observed in rodent cell lines?

DHFR GENE AMPLIFICATION IN CULTURED HUMAN AND RODENT CELLS FOLLOWING METHOTREXATE SELECTION WITH AND WITHOUT GENOTOXIC PRETREATMENT

Schimke et al. (1978) were the first to report that resistance to the metabolic inhibitor methotrexate (MTX), which is used in tumor therapy, is associated with the amplification of the dihydrofolate reductase (DHFR) gene in cultured mouse cells. Many similar observations followed (Wahl et al., 1979; Bostock and Tyler-Smith, 1982; Andrullis and Siminovitch, 1982), all showing that resistance to an inhibitor of a specific enzyme is the result of amplification of the gene coding for this enzyme. Most of the data published have been obtained with rodent cells. Hamster cells in particular amplify large segments of DNA (Guilotto et al., 1986; Looney and Hamlin, 1987).

Influence of Genomic Instability on Gene Amplification

In order to answer the question of whether gene amplification is one of the initial events in the multistep process of human carcinogenesis, we studied this phenomenon in human diploid skin fibroblasts. Our efforts to induce DHFR gene amplification in these cells by stepwise increasing the concentration of MTX in the culture medium were without success for years. Human primary skin fibroblasts of various origin and two human fibroblast cell lines transformed with SV40 (healthy donors: GM637, MRC5CVI) did not amplify their DHFR genes to any measurable extent and were unable to grow in high MTX concentrations, irrespective of prior treatment with DNA damaging agents.

However, using the same protocol (experimental details have been published in Lücke-Huhle et al., 1987), we could induce gene amplification in SV40-transformed skin fibroblasts of a patient with the hereditary disorder ataxia telangiectasia (AT). AT patients suffer from X-ray sensitivity and increased cancer risk (Taylor et al., 1975), and the instability of their cellular genome has been documented karyotypically (Lehman, 1982). Initially, AT cells (AT5BI-VA) are as sensitive to MTX as cells from healthy donors, showing ED50 values between 0.01 to 0.04 μM MTX (ED50 is the effective dose of MTX, causing 50% growth inhibition in comparison to control cells). After selection over a period of months, MTX-resistant (MTX^r) subclones of AT cells were obtained. They were able to grow in MTX concentrations of up to 100 μM, that is, four orders of magnitude higher than the concentration the starting cells could survive. A detailed analysis of the genomic organization revealed amplification of all exons of the DHFR structural gene, but no amplification of pseudogenes (Fig. 1A). The value of 12 DHFR gene copies per cell genome (corresponding to a sixfold amplification) in the resistant AT cells (Table 1) might therefore be much higher if the high background of unamplified pseudogenes were subtracted.

The genomic instability of AT cells is reflected not only by DHFR amplification but also by the concomitant loss of nearly all SV40 sequences (Fig. 1B) during the MTX selection procedure (originally there are 10 SV40 copies integrated per AT5BI-VA genome).

Pretreatment with a number of genotoxic agents, including radiation, chemical carcinogens, hypoxia, or metabolic inhibitors of DNA synthesis, increases the frequency of methotrexate resistance in mouse and hamster cells as a result of DHFR gene amplification (Schimke, 1984a). γ-Radiation pretreatment of AT cells also increased the copy number of DHFR genes by an additional factor of two (Fig. 2, ATX 10 cells) and also led to a fourfold amplification of the cellular oncogene Ki-ras. The product of Ki-ras might support growth of the human fibroblasts after loss of the SV40 sequences. Thus, both, MTX- and DNA-damage-induced amplification is fascilitated in AT cells in comparison to normal human fibroblasts.

Influence of Cell Differentiation on Gene Amplification

The influence of differentiation on the ability of a cell to amplify a specific DNA sequence has been studied in F9 mouse teratocarcinoma

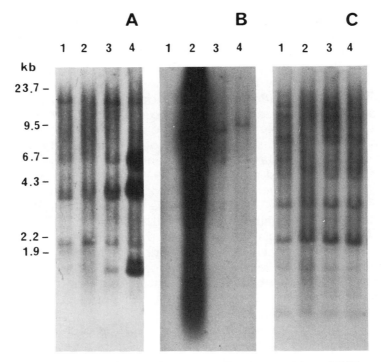

FIGURE 1. Amplification of the structural DHFR gene fragments and loss of SV40 sequences following irradiation and MTX selection is shown in the hybridization pattern of Eco RI digested DNAs (10 μg each) from AT5BI primary fibroblasts (lane 1), SV40-transformed AT5BI-VA (lane 2), γ-irradiated AT5BI-VA (lane 3), and MTX-resistant ataxia cells (ATX10: lane 4). The Southern blot was hybridized either to ^{32}P-labeled DHFR (A), SV40 DNA (B), or an α-actin probe (C).

TABLE 1. Compilation of Data on DHFR Amplification Following Methotrexate Selection with and without Radiation Pretreatment

Cell line	Species	DHFR amplification	Pretreatment
AT5BI-VA	Human	6- to 12-fold	Without, γ-rays, α-particles
GM637	Human	None	γ-Rays, α-particles, UV
MRC5CVI	Human	None	γ-Rays, α-particles, UV
F9	Mouse	43-fold	Without, γ-rays
F9-RA	Mouse	None	Without, γ-rays
BSp73 ASML	Rat	13-fold	Without
BSp73 AS	Rat	None	Without

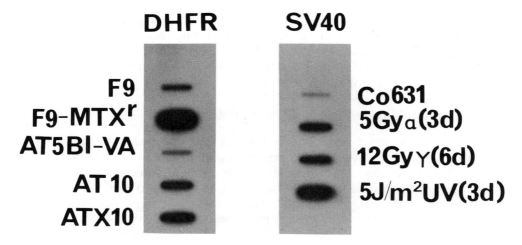

DHFR **SV40**

F9

F9-MTX^r

AT5BI-VA

AT 10

ATX10

Co631

5Gy$_\alpha$(3d)

12Gy$_\gamma$(6d)

5J/m^{2}UV(3d)

FIGURE 2. Autoradiograms of DNA slot blots (1 μg each) hybridized with ^{32}P-labeled DHFR or SV40 DNA probe show DHFR amplification in F9-MTX^r cell clones as well as in AT 10 (MTX^r ataxia cell line) and in ATX 10 (MTX^r ataxia cell line pretreated with γ-irradiation). SV40 DNA hybridizations show maximum values of SV40 amplification in Co631 cells following various treatments.

stem cells. These cells differentiate easily into an early embryonic cell type known as parietal endoderm in response to retinoic acid (given at a dose of 3.3 × 10^{-7} M for at least 6 d). Differentiation was documented by morphology changes, cell cycle prolongation, and reduced expression of the cellular oncogene *c-myc*. Initially (prior to MTX contact), nondifferentiated F9 teratocarcinoma cells and differentiated F9-RA endodermal cells are equally sensitive to MTX (ED50 = 0.02 μM). Exposure of exponentially growing cultures to progressively increasing concentrations of MTX yielded highly resistant cells within 4 mo, but only from the nondifferentiated F9 cultures.

While the ED50 value of resistant cells was 1000 μM (i.e., five orders of magnitude more resistant than the starting culture), the isogenic but differentiated F9-RA cells remained sensitive (tolerating only 0.1 μM). Apparently differentiation prevented adaptation to MTX resistance (Lücke-Huhle and Herrlich, 1990).

Resistance of the F9-MTX^r cells was associated with the acquisition of additional DHFR gene copies (Fig. 2 and Table 1: a maximum of 85 copies per cell was found in the cells resistant to 1000 μM MTX) and with increased DHFR mRNA levels.

Exposure to 5 Gy of ^{60}Co γ-irradiation prior to MTX selection enhanced amplification in the undifferentiated F9 cells, whereas no amplification could be detected in the differentiated F9-RA cells under all conditions tested.

Undifferentiated F9 cells not only amplified easily, their MTX^r cell clones also quickly lost their extra DHFR gene copies in the absence of MTX selection (95% within 3 mo). Interestingly, this loss could be prevented by retinoic acid treatment. Thus, it appears that differentiation of cells by retinoic acid affects both induced gene amplification and noninduced mobilization of DNA.

Gene Amplification and Malignant Progression

Clinical investigations have shown a correlation between the number of positive nodes in certain human tumor patients and the existence of amplified oncogenes in tumor cells. This might make amplification a reliable prognostic parameter (Slamon et al., 1989). To find out whether tumor cells with a high probability for metastatic progression also have a higher potential to amplify genes under selective pressure in cell culture, we examined a pair of tumor cell lines (BSp73 AS and BSp73 ASML) derived from a spontaneous rat pancreatic adenocarcinoma (BSp73: Matzku et al., 1983). BSp73 cells remain localized upon subcutaneous injection, while BSp73 ASML cells (originating from a lung metastasis) are highly metastatic if injected into isogenic rats. Exposure to stepwise increasing concentrations of MTX resulted in methotrexate-resistant ASML cells. Their ability to grow in 100 μM MTX was mainly due to a 13-fold amplification of their DHFR gene (Table 1). The nonmetastasizing cells, however, within the same short period of time did not show DHFR gene amplification and were unable to grow in medium containing high MTX concentrations (Lücke-Huhle, unpublished). These results may simply reflect an increase in genomic instability with tumor progression. However, the consequences may specifically contribute to the metastatic behavior, in that amplified DNA will partly be reintegrated and thus will cause mutations in various genes from which the metastatic properties could be selected in vivo.

SIMIAN VIRUS 40 SEQUENCES AS INDICATOR GENE FOR SELECTIVE DNA AMPLIFICATION

To study gene amplification directly after exposure of cells to carcinogenic agents rather than after the long time needed for establishing cell lines containing amplified genes, we used an experimental model system. Co631 cells (constructed by Lavi, 1981) are simian virus 40 wild type (SV40) transformed Chinese hamster embryo cells that contain about five viral copies integrated per cellular genome. The integrated SV40 sequences are used as an endogenous indicator gene, with the advantage that changes in gene copy number occur in the majority of cells and therefore can be monitored directly following carcinogen treatment.

A large number of carcinogenic agents have been shown to induce SV40 amplification in Co631 cells: dimethylbenzanthracene (DMBA),

N-methyl-N-nitro-N-nitrosoguanidine (MNNG), benzopyrenes, aflatoxins, phenols, various chemotherapeutic agents, and all types of radiation (Lavi and Etkin, 1981; Lücke-Huhle et al., 1986; Ehrfeld et al., 1986; Pool et al., 1989). The methodology employed in these experiments has been described in detail elsewhere (Lücke-Huhle et al., 1986). In short, cells were exposed as monolayers to the carcinogen and analyzed for changes in gene copy number by dispersed cell blotting or DNA blotting in combination with highly sensitive DNA hybridization methods at various times after treatment (1–7 d).

Upon exposure of Co631 cells to various types of external (^{60}Co γ-rays, ^{241}Am α-particles, UV) or internal radiation (caused by the decay of ^{125}I incorporated into DNA in form of [5-^{125}I]iododeoxyuridine, I-UdR), SV40 was amplified (Fig. 2) without leading to the production of intact virus. The degree of amplification depended on dose and time after irradiation. Amplification was already well detectable at d 1 (Fig. 3). In Table 2 maximum effects of various doses have been compiled. The time point of maximum SV40 amplification was d 3 after alpha particles, UV exposure, or chemical carcinogens, and d 6–7 after γ-ray exposure.

In the case of ^{125}I, SV40 amplification continued to increase through d 7 (the latest time point investigated) due to radiation emitted by the decay of the incorporated ^{125}I. In all cases of radiation treatment, the increase in copy number followed an exponential kinetics (Fig. 3), indicating that all copies (including the amplified ones) participate in rereplication. This finding excludes an extrachromosomal rolling circle mechanism as well as pure recombination events as mechanisms of SV40 amplification.

SV40 amplification is a transient process. The number of amplified copies decreased by d 4 or 7 after radiation (Fig. 3) as well as after exposure to chemicals (Lavi, 1982). This time course is independent of the dose of carcinogen and might be explained by the fact that amplified SV40 sequences, not offering any growth advantage to the host, get lost during cell divisions.

Direct measurement of induced DNA amplification is not restricted to integrated SV40 sequences in Chinese hamster embryo cells. Similar observations have been made with polyoma virus-transformed rat cells (Lambert et al., 1986).

As mentioned before, SV40 amplification is dose-dependent. Increasing doses of radiation induce higher enhancement of SV40 amplification but concomitantly reduce cell survival. Because it is, of course, the surviving fraction of the treated cells that is of interest with respect to mutagenic and carcinogenic risk, it had to be excluded that gene amplification is simply the response of dying cells. In situ hybridization of surviving colonies (Rossman and Rubin, 1988) and a comparison of their hybridization signal with that of untreated colonies of equal cell number (Lücke-Huhle et al., 1990b) demonstrated that all surviving clones had

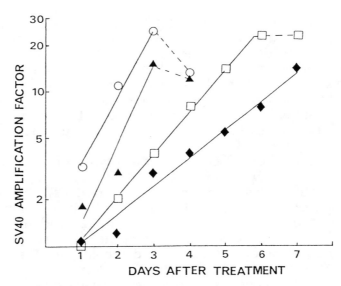

FIGURE 3. Exponential increase in SV40 DNA amplification with time after irradiation: ○ 5 J/m² of UV; ▲ 5 Gy of ²⁴¹Am α-particles; □ 12 Gy of ⁶⁰Co γ-rays; ◆ 1.85 kBq/ml ¹²⁵IUdR.

TABLE 2. Maximum SV40 Amplification in Chinese Hamster Embryo Cells (Co631) after Treatment with Various DNA-Damaging Agents and/or the Metabolic Inhibitor Cycloheximide (CHX)

Agent	Dose	Cell kill	SV40 amplification
⁶⁰Co γ	2.5 Gy	20%	5×
	8.0 Gy	90%	13×
	12.0 Gy	98%	23×
UV	2 J/m²	50%	3×
	5 J/m²	95%	25×
	15 J/m²	>99%	4×
²⁴¹Am α	0.7 Gy	50%	2×
	2.5 Gy	94%	5×
	5.0 Gy	>99%	15×
¹²⁵IUdR	1.6 × 10⁻² pCi/cell	20%	2×
	4.8 × 10⁻² pCi/cell	75%	3×
	8.0 × 10⁻² pCi/cell	93%	13×
DMBA	5 μg/ml	nd[b]	11×
MNNG	5 μg/ml	nd	25×
CHX	0.5 μg/ml	24%	8× (170×)[a]
	1.0 μg/ml	44%	18× (360×)[a]
	10.0 μg/ml	97%	10× (130×)[a]

[a]In combination with 1.5 J/m² UV.
[b]Not determined.

amplified their SV40 sequences as efficiently as the abortive colonies that had lost their capacity for sustained cell proliferation.

A real limitation in the dose-dependent increase in SV40 amplification is enforced by the density of damaged sites. Therefore, a dose of 15 J/m^2 induced much less SV40 amplification than 5 J/m^2 (Table 2). Based on the calculations of Setlow (1967) and Wulff (1963), at 5 J/m^2 (the dose of maximum effect) about 3 thymidine dimers are produced within the five integrated SV40 copies per cell. A still higher number of lesions produced with higher doses of UV seems to interfere with the amplification process by damaging the amplicon itself, even if the lesions are repairable at a later time.

Amplification-inducing lesions are not identical with killing lesions. That is why survivors and nonsurvivors amplified about equally well after treatment with DNA-damaging agents and why different types of radiation (sparsely ionizing ^{60}Co γ-rays, densely ionizing ^{241}Am α-particles, Auger electrons originating from the decay of ^{125}I or UV radiation) that cause completely different spectra of DNA damage (single-strand breaks, double-strand breaks, various kinds of base damage) and have different killing efficiencies were about equally effective in inducing amplification. For instance, a dose of 2.5 Gy of γ- or α-rays yielded a fivefold SV40 amplification, although the killing efficiency of alpha particles is fivefold higher (Table 2).

The densely ionizing radiation emitted by the decay of the incorporated ^{125}I atoms induced (in addition to SV40 amplification) a twofold amplification of the cellular oncogenes Ki-ras and Ha-ras, the genes that have also been found amplified after treatment with chemicals and in radiation-induced animal tumors (Guerrero et al., 1984; Barbacid, 1986).

INITIATING EVENTS OF SV40 AMPLIFICATION INDUCED BY RADIATION AND OTHER DNA-DAMAGING AGENTS

We and others have shown that the process of SV40 amplification requires an intact origin of replication and a functional large T antigen (Lücke-Huhle and Herrlich, 1987; Lavi, 1982). A cell line with defective origin of replication (C 812) did not show SV40 amplification after irradiation or treatment with chemical carcinogens. The importance of an active large T antigen was demonstrated in experiments using a thermosensitive T-antigen mutant (C 1102 ts). Amplification then occurred only at the permissive temperature of 33°C when the T antigen produced was stable.

The fact that amplification has been found at doses of alpha particles as low as 0.7 Gy (a dose too low to hit the SV40 sequences directly in a large number of cells) suggested already that the site of the inducing lesion and the site of amplification were different entities and that these entities communicated by a transacting mechanism. The existence of a transacting cellular factor was confirmed by cell fusion experiments between irradiated or carcinogen-treated cells and untreated cells carrying the indicator gene (Lücke-Huhle and Herrlich, 1987; Nomura and Oishi, 1984; Lambert et al., 1986; Ronai and Weinstein, 1988; Kleinberger et al., 1988). Amplification of the indicator gene in the heterokaryon indicates the transfer of a signal through the cytoplasm by one or more cellular factors that has been induced or activated by DNA damage.

The postulate of a transacting factor is supported by the successful induction of SV40 amplification by the indirect method of introducing damaged DNA (by UV or MNNG treatment) of oligonucleotide size into Co631 cells (Mai et al., 1990).

Because overreplication requires and starts at the SV40 origin of replication, it was supposed that the inducing factor acts at this site. By DNase I footprinting technique, the binding of a protein in cell extracts from UV-treated cells was detected at position 5207–5225, referred to as the early domain of the minimal origin of replication of SV40 (Lücke-Huhle et al., 1990a). Using a synthetic oligonucleotide (Table 3) comprising this domain for gel retardation experiments, the specific binding of a cellular protein was detected. Binding activity was enhanced in extracts of irradiated or chemically treated cells (γ-radiation, α-radiation, UV, MNNG). A dose of 5 J/m^2 of UV enhanced binding activity at least 10-fold within 2 min (Mai, 1991) and activity persisted for several hours. Point mutations of the binding motif decreased the binding efficiency in vitro considerably. The specificity of the binding was shown by competition assays with unlabeled oligonucleotides and by using protein extracts from various cell lines (HeLa, NIH3T3, CHO9) after irradiation or treatment with MNNG (Lücke-Huhle et al., 1989).

The increase in binding activity did not require new protein synthesis. Posttranslational activation has been suggested by the fact that binding of the factor cannot be blocked by excessive doses of the protein inhibitors anisomycin or cycloheximide. But treatment with alkaline phosphatase eliminated binding to the early domain, whereas the enzyme in the presence of the phosphatase inhibitor molybdate (4 mM) did not. Therefore, phosphorylation of one or more specific sites seems to be necessary for the binding of this cellular protein to DNA (Mai et al., 1990).

Treatment with anisomycin and cycloheximide not only did not prevent binding to the oligonucleotide comprising the early domain, but itself caused elevated activity and showed superadditive effects in combination with radiation or MNNG.

This specific binding activity is also present in cell lines not containing SV40 sequences and is expressed at a higher constitutive level, without carcinogenic treatment, in permissive cell lines such as monkey kidney cells (CVI) and human skin fibroblasts (Lücke-Huhle, unpublished; Traut and Fanning, 1988).

TABLE 3. Inhibition of UV-Induced SV40 DNA Amplification by In Vivo Competition for the "Early Domain" Binding Protein with Various Types of Oligonucleotides

Type of oligonucleotide		UV-induced SV40 amplification
Without		15×
	5207 5225	
WT	5' GATCCTCCTCACTACTTCTGGAATG 3'	
WT	3' GAGGAGTGATGAAGACCTTACCTAG 5'	1.1×
	* *	
M1	5' GATCCTCCTCCCTCCTTCTGGAATG 3'	
M2	3' GAGGAGGGAGGAAGACCTTACCTAG 5'	8×
	* *	
M3	5' GATCCTCCTCGCGACTTCTGGAATG 3'	
M4	3' GAGGAGCGCTGAAGACCTTACCTAG 5'	10×
	*	
M5	5' GATCCTCCTCACTACTTTTGGAATG 3'	
M6	3' GAGGAGTGATGAAGACCTTACCTAG 5'	8×
SYN16/1	5' GATCTAGAAAATTATTAACCTCA 3'	
SYN16/2	3' ATCTTTTAATAATTGGAGTTCGA 5'	10×

minimal origin

T-AG.I   early d   T-AG.II   A/T   AP1   (21bp)$_3$ (72bp)$_2$

Note. The cells were analyzed for their amount of SV40 copies 3 d after irradiation with 5 J/m^2 UV and incubation with and without the addition of synthetic double-stranded oligonucleotides (WT early domain wild type, M$_1$/M$_2$, M$_3$/M$_4$, and M$_5$/M$_6$ mutant sequences of the early domain, SYN16 completely unrelated sequence). Amplification factors are mean values of four to eight experiments each.

    The binding of one or more cellular factors to the early domain is absolutely necessary as the initial step in the process of DNA amplification. In an in vivo assay an excess of a double-stranded oligonucleotide comprising the early domain from position 5207 to 5225, added directly after carcinogenic treatment, totally blocked radiation- as well as other damage-induced SV40 amplification (Table 3). Oligonucleotides of small size (in our case 25 base pairs) are taken up by the cells from the culture medium if added in high concentration (30 μM). Thus, we suppose that the competition experiment in vivo and the binding in vitro shown by gel bandshifts probably concern the same protein. Mutants of the early domain sequence and unrelated sequences did not interfere with amplification even if added in 50 μM concentration.

    Induced amplification in vivo was as resistant to inhibitors of protein synthesis as increased binding activity in the gel shift experiments. In fact, an even more drastic increase in SV40 amplification occurred following the combination of UV and cycloheximide (Table 2). How cycloheximide or anisomycin affects activity is not yet clear. Cycloheximide could cause cellular deprivation of a labile regulator affecting transcriptional rate, mRNA half-life, or translation. Carcinogens might act through another mechanism (e.g., activation of a positive factor). Otherwise the additivity is difficult to explain.

Characterization of the early domain binding complex by UV cross-linking and SDS polyacrylamide gel electrophoresis showed three proteins of 67, 64, and 46 kD, but only the binding activity of the first two proteins was inducible by DNA-damaging agents (Mai et al., 1990).

Reports on increased SV40 replication in Myc-expressing human lymphoma cells (Classon et al., 1987), as well as the result by Ariga et al. (1989) describing an origin of replication for c-myc that shows homology to the SV40 early domain and to which Myc protein binds itself, prompted us to study the involvement of c-myc protein in gene amplification. Using antibodies against Myc, a specific interference with the formation of the gel retardation complex was found, suggesting that the formation of the DNA-protein complex, presumably essential for SV40 amplification, is or at least requires Myc. These results are strongly supported by stable Co631 transfectants carrying a construct with the c-myc c-DNA under inducible MMTV promoter control. Overexpression of c-myc increases the early domain binding activity and induces SV40 amplification without existing DNA damage. Overexpression of c-myc also sufficed to induce a three- to fourfold DHFR amplification (Mai et al., 1990) in these cells.

CONSEQUENCES OF AMPLIFIED DNA SEQUENCES FOR SURVIVING CELLS

Reintegration of Excised SV40 DNA Copies at New Genomic Sites

By Southern hybridization the number and fate of DNA sequences can be examined. The five genomically integrated SV40 copies of untreated Co631 cells yield five characteristic bands after Xba I digestion. Xba I is an enzyme that does not cut within SV40 sequences. This fragment pattern is reproducible for all individual clones selected prior to treatment. Thus, the sequences do not spontaneously rearrange or amplify. After treatment, the amplified SV40 copies in surviving cells get lost within the following 20–30 cell generations. In 12 out of 36 clones, however, the remaining copies showed, in addition to the integration sites of the control cells, new fragments that must have occurred as a result of the radiation-induced amplification (Lücke-Huhle et al., 1990b). This finding, as well as results by others (Steinberg et al., 1989), suggests that radiation or chemical carcinogens can generate mutations by inducing overreplication of chromosome segments that are then substrates of enzymatic rearrangements. DNA rearrangements by chromosomal transposition represent a particularly intriguing mechanism, because it could bring oncogenes under cis control of a highly active chromosome region.

Increased Frequency of Chromosome Aberrations and SCEs in AT Cells with Amplified DHFR Genes

While skin fibroblasts of healthy human donors do not seem to be able to adapt to higher concentrations of MTX, AT cells readily underwent amplification of their DHFR gene and were able to grow in MTX concentrations up to 100 μM. At the chromosome level, untreated AT cells already show increased numbers of spontaneous chromosomal aberrations and an enhanced response to induction of aberrations by ionizing radiation (Lehman, 1982). These features are indicative of the genomic instability of AT cells and made them especially interesting for the investigation of gene amplification.

After SV40 transformation of primary AT skin fibroblasts, the chromosome number increased from 46 to 63 chromosomes (Table 4), while in the MTX^r clones the chromosome numbers reduced again. It is unclear whether this finding is responsible for the observed loss of SV40 sequences in MTX^r clones (Fig. 1).

Sister chromatid exchange (SCE) frequencies (an indicator of DNA damage and its intrachromosomal recombination) are slightly increased after SV40 transformation. However, SCE frequency is significantly higher in the MTX^r ATX 10 line. An augmentation of the genomic instability by acquisition of amplified gene copies was also revealed by an obvious rise in chromosome rearrangements measured as exchange figures and ring chromosomes (defined as aberrations in Table 4) in the MTX^r ATX 10 cell line (Speit and Lücke-Huhle, 1987).

TABLE 4. Cytogenetic Characteristics of Ataxia Telangiectasia Cell Lines After SV40 Transformation and Selection for MTX Resistance

Cell line	Chromosomes/ metaphase[a]	Polyploid cells (%)[b]	Metaphases with aberrations (%)[c]	SCEs/ chromosome
AT5BI	46	5	3	0.15
AT5BI-VA	63	13	4	0.21
ATXO	61	21	10	0.32
ATX10	57	8	18	0.44

[a]Mean of 40 cells.
[b]>90 Chromosomes per metaphase.
[c]100 Metaphases were analyzed per culture.

CONCLUSIONS

Due to their greater karyotypic stability, the induction of gene amplification in normal human cells is more difficult to achieve than in rodent cells, tumor cell lines, or human cell lines originating from patients suffering from a repair deficiency syndrome. This correlates with the findings on neoplastic transformation in vitro.

The property of easily amplifying and losing DNA fragments may be one of the mechanisms leading to the emergence of clonal subpopulations with continuously increasing malignant behavior and thus may explain the higher cancer incidence (e.g., in AT patients) and the increasing capacity for gene amplification with increasing metastatic potential. Amplified DNA is a source of genomic changes causing mutant phenotypes in mammalian cells by large-scale alterations in DNA structure as well as localized changes in sequences such as arise by SCEs or new integration sites of amplicons.

The finding that F9 teratocarcinoma stem cells are more susceptible to DHFR amplification than the isogenic differentiated F9-RA endodermal cells suggests that loss of differentiation plays a role in cell sensitivity to amplification and transformation. Immortalized cell lines and tumor cells may have regained the stem-cell property rendering them amplification proficient.

The great importance of the *c-myc* protein was shown by its involvement in the formation of an origin complex within the SV40 minimal origin of replication. Its exact function and role in the amplification and replication process need to be explored.

REFERENCES

Anderson, R. P., and Roth, J. R. 1977. Tandem genetic duplications in phage and bacteria. *Annu. Rev. Microbiol.* 31:473–505.

Andrullis, J. L., and Siminovitch, L. 1982. Amplification of the gene for asparagine synthetase. In *Gene Amplification,* ed. R. T. Schimke, pp. 75–80. Cold Spring Harbor, N.Y.: Cold Spring Harbor Laboratory.

Ariga, H., Imamura, Y., and Iguchi-Ariga, S. M. M. 1989. DNA replication origin and transcriptional enhancer in *c-myc* gene share the c-myc protein-binding sequences. *EMBO J.* 8:4273–4279.

Balaban-Malenbaum, G., and Gilbert, F. 1977. Double minute chromosomes and the homogeneously staining regions in chromosomes of a human neuroblastoma cell line. *Science* 198:739–741.

Barbacid, M. 1986. Oncogenes and human cancer: Cause or consequence? *Carcinogenesis* 7:1037–1042.

Biedler, J. L., and Spengler, B. A. 1976. Methaphase chromosome anomaly: Association with drug resistance and cell-specific products. *Science* 191:185–189.

Bostock, C. J., and Tyler-Smith, C. 1982. Amplification of dihydrofolate reductase genes and other DNA sequences in mouse cells. In *Gene Amplification,* ed. R. T. Schimke, pp. 15–21. Cold Spring Harbor, N.Y.: Cold Spring Harbor Laboratory.

Classon, M., Henriksson, M., Klein, G., and Hammaskjold, M.-L. 1987. Elevated *c-myc* expression facilitates the replication of SV40 DNA in human lymphoma cells. *Nature* 30:272–274.

DeCicco, D. V., and Spradling, A. D. 1984. Localization of a cis-acting element responsible for the developmentally regulated amplification of *Drosophila* chorion genes. *Cell* 38:45–54.

Ehrfeld, A., Planas Bohne, F., and Lücke-Huhle, C. 1986. Amplification of oncogenes and integrated SV40 sequences in mammalian cells by the decay of incorporated iodine-125. *Rad. Res.* 108:43–51.

Guerrero, I., Villasante, A., Carces, V., and Pellicer, A. 1984. Activation of a *c-Ki-ras* oncogene by somatic mutation in mouse lymphomas induced by gamma-radiation. *Science* 225:1159–1162.

Guilotto, E. O., Saito, J., and Stark, G. R. 1986. Structure of DNA formed in the first step of CAD gene amplification. *EMBO J.* 5:2114–2121.

Kleinberger, T., Flint, Y. B., Blank, M., Etkin, S., and Lavi, S. 1988. Carcinogen-induced trans activation of gene expression. *Mol. Cell. Biol.* 8:1366–1370.

Lambert, M. E., Pellegrini, S., Gattoni-Celli, S., and Weinstein, I. B. 1986. Carcinogen induced asynchronous replication of polyoma DNA is mediated by a trans-acting factor. *Carcinogenesis* 7:1011–1017.

Lavi, S. 1981. Carcinogen-mediated amplification of viral DNA sequences in simian virus 40-transformed Chinese hamster embryo cells. *Proc. Natl. Acad. Sci. USA* 78:6144–6148.

Lavi, S. 1982. Carcinogen-mediated amplification of specific DNA sequences. *J. Cell. Biochem.* 18:149–156.

Lavi, S., and Etkin, S. 1981. Carcinogen-mediated induction by SV40 DNA synthesis in SV40 transformed Chinese hamster embryo cells. *Carcinogenesis* 2:417–423.

Lehman, A. R. 1982. The cellular and molecular responses of ataxia telangiectasia cells to DNA damage. In *Ataxia-Telangiectasia,* eds. B. A. Bridges and D. G. Harnden, pp. 83–102. New York: Wiley.

Long, E. O., and Dawid, I. B. 1980. Repeated genes in eukaryotes. *Annu. Rev. Biochem.* 49:727–764.

Looney, J. E., and Hamlin, J. L. 1987. Isolation of the amplified dihydrofolate reductase domain from methotrexate-resistant Chinese hamster ovary cells. *Mol. Cell. Biol.* 7:569–577.

Lücke-Huhle, C., Gloss, B., and Herrlich, P. 1990a. Radiation-induced gene amplification in rodent and human cells. *Acta Biol. Hung.* 41:159–171.

Lücke-Huhle, C., and Herrlich, P. 1986. Gene amplification in mammalian cells after exposure to ionizing radiation and UV. In *Radiation Carcinogenesis and DNA Alterations,* eds. F. J. Burns, A. C. Upton, and G. Silini, pp. 405–411. Amsterdam: Plenum Press.

Lücke-Huhle, C., and Herrlich, P. 1987. Alpha-radiation-induced amplification of integrated SV40 sequences is mediated by a trans-acting mechanism. *Int. J. Cancer* 39:94–98.

Lücke-Huhle, C., and Herrlich, P. 1991. Retinoic acid induced differentiation prevents gene amplification in teratocarcinoma stem cells. *Int. J. Cancer* 47:461–465.

Lücke-Huhle, C., Hinrichs, S., and Speit, G. 1987. DHFR gene amplification in cultured skin fibroblasts of ataxia telangiectasia patients after methotrexate selection. *Carcinogenesis* 8:1801–1806.

Lücke-Huhle, C., Mai, S., and Herrlich, P. 1989. UV induced early-domain binding factor as the limiting component of simian virus 40 DNA amplification in rodent cells. *Mol. Cell. Biol.* 9:4812–4818.

Lücke-Huhle, C., Pech, M., and Herrlich, P. 1986. Selective gene amplification in mammalian cells after exposure to [60]Co γ rays, [241]Am α particles, or UV light. *Radiat. Res.* 106:345–355.

Lücke-Huhle, C., Pech, M., and Herrlich, P. 1990b. SV40 DNA amplification and reintegration in surviving hamster cells after [60]Co γ-irradiation. *Int. J. Radiat. Biol.* 58:577–588.

Mai, S. 1991. Mechanismen der Mutagen-induzierten zellulären Streßreaktion: von DNA Schädigung zu c-Myc abhängiger Genamplifikation. Dissertationsschrift, Universität Karlsruhe.

Mai, S., Lücke-Huhle, C., Kaina, B., Rahmsdorf, H. J., Stein, B., Ponta, H., and Herrlich, P. 1990. Ionizing radiation induced formation of a replication origin binding complex involving the product of the cellular oncogene c-myc. In *UCLA Symposia on Molecular and Cellular Biology, Ionizing Radiation Damage to DNA: Molecular Aspects,* eds. S. Wallace and R. Painter, pp. 319–331. Philadelphia: Wiley/Liss.

Matzku, S., Komitowski, D., Mildenberger, M., and Zöller, M. 1983. Characterization of Bsp 73, a spontaneous rat tumor and its in vivo selected variants showing different metastasizing capacities. *Invest. Methods* 3:109–123.

Nomura, S., and Oishi, M. 1984. UV-irradiation induces an activity which stimulates simian virus 40 rescue upon cell fusion. *Mol. Cell. Biol.* 4:1159–1162.

Pool, B. L., Yalkinoglu, A. Ö., Klein, P., and Schlehofer, J. R. 1989. DNA amplification in genetic toxicology. *Mutat. Res.* 213:61–72.

Ronai, Z. A., and Weinstein, I. B. 1988. Identification of a UV-induced trans-acting protein that stimulates polyomavirus DNA replication. *J. Virol.* 62:1057–1060.

Rossman, T. G., and Rubin, L. M. 1988. Colony hybridization to identify mammalian cells containing amplified, transfected, or expressed sequences. *Som. Cell Molec. Genet.* 14:321–328.

Schimke, R. T. 1984a. Gene amplification in cultured animal cells. *Cell* 37:705–713.

Schimke, R. T. 1984b. Gene amplification, drug resistance, and cancer. *Cancer Res.* 44:1735–1742.

Schimke, R. T., Kaufman, R. J., Alt, F. W., and Kellems, R. F. 1978. Gene amplification and drug resistance in cultured murine cells. *Science* 202:1051–1055.

Schwab, M., and Amler, L. C. 1990. Amplification of cellular oncogenes. A predictor of clinical outcome in human cancer. *Genes Chromosomes Cancer* 1:181–193.

Setlow, J. K. 1967. The effects of ultraviolet radiation and photoreactivation. In *Comprehensive Biochemistry*, vol. 27, *Photobiology, Ionizing Radiations*, eds. M. Florkin and E. M. Stotz, pp. 176–178. New York: Elsevier.

Shah, D. M., Horsch, R. B., Klee, H. J., Kishore, G. M., Winter, J. A., Turner, N. E., Hironaka, C. M., Sanders, P. R., Gasser, S., Arykent, S., Siegel, N. R., Rogers, S. G., and Fowley, R. T. 1986. Engineering herbicide tolerance in transgenic plants. *Science* 233:478–481.

Slamon, D. J., Godolphin, W., Jones, L. A., Holt, J. A., Wong, S. G., Keith, D. E., Levin, W. J., Stuart, S. G., Udove, J., Ullrich, A., and Press, M. F. 1989. Studies of the HER-2/neu proto-oncogene in human breast and ovarian cancer. *Science* 244:707–712.

Speit, G., and Lücke-Huhle, C. 1987. Chromosomal changes associated with methotrexate-induced gene amplification in ataxia telangiectasia cells. *Ann. Univ. Sarav. Med. Suppl.* 7:302–304.

Spradling, A. C., and Mahowald, A. P. 1980. Amplification of genes for chorion proteins during oogenesis in *Drosophila melanogaster. Proc. Natl. Acad. Sci. USA* 7:1069–1100.

Stark, G. R., Debatisse, M., Giulotto, E., and Wahl, G. M. 1989. Recent progress in understanding mechanisms of mammalian DNA amplification. *Cell* 57:901–908.

Stark, G. R., and Wahl, G. M. 1984. Gene amplification. *Annu. Rev. Biochem.* 53:447–492.

Steinberg, M. L., Rossman, T. G., Morris, A., and Chen, G. 1989. Specific high frequency rearrangements induced by MNNG in SV40-infected human keratinocytes. *Carcinogenesis* 10:1801–1807.

Taylor, A. M. R., Harnden, D. G., Arlett, C. F., Harcourt, S. A., Lehman, A. R., Stevens, S., and Bridges, B. A. 1975. Ataxia telangiectasia: A human mutation with abnormal radiation sensitivity. *Nature* 258:427–429.

Traut, W., and Fanning, E. 1988. Sequence-specific interaction between a cellular DNA-binding protein and the simian virus 40 origin of DNA replication. *Mol. Cell. Biol.* 8:903–911.

Wahl, G. M., Padgett, R. M., and Stark, G. R. 1979. Gene amplification causes over production of the first three enzymes of UMP synthesis in *N*(phosphoacetyl 1-aspartate)-resistant hamster cells. *J. Biol. Chem.* 254:8679–8689.

Wulff, D. L. 1963. Kinetic of thymine photodimerization in DNA. *Biophys. J.* 3:355–362.

Zur Hausen, H., and Schlehofer, J. R. (eds.). 1987. The role of DNA amplification in carcinogenesis. In *Accomplishments in Oncology*, vol. 2, no. 1. Philadelphia: Lippincott.

7 | CARCINOGEN-INDUCIBLE PROTEINS AS MEDIATORS OF ASYNCHRONOUS REPLICATION OF DNA TUMOR VIRUSES

Zeev Ronai, Susan Rutberg, Judy Hammelburger, Richard Robinson

Molecular Carcinogenesis Program, American Health Foundation, Valhalla, New York

INTRODUCTION

The response of DNA tumor viruses to carcinogens is a well-documented phenomenon involving asynchronous DNA replication (also referred to as selective amplification). This response has been demonstrated for a variety of DNA tumor viruses including SV40, polyoma, herpes, and adeno-associated helper virus. The use of molecular biology methodologies has enabled a better understanding of this type of DNA replication at the molecular level and has shed new light on the use of viral DNA sequences as a marker system for examining the effects of carcinogens on mammalian cells. The response of DNA tumor viruses provides an attractive system for studying the effects of carcinogens on both viral and cellular genes, as certain cellular genes are known to carry regulatory domains that have sequences in common with the regulatory regions of these viruses. Thus, the use of such a model system can add substantial knowledge to our understanding of the responses of cellular genes to DNA damaging agents. Asynchronous replication of viral DNA sequences may also provide a useful system for understanding mechanisms underlying disorders in DNA replication during the carcinogenesis process. In addition, studies involving viral replication may provide insights into mechanisms by which physical or chemical carcinogens can act synergistically with certain viruses to enhance cell transformation. Because certain human tumors may be caused by the combined effects of viruses and carcinogen damage, this system could be of direct relevance to the origin of certain human tumors (Weinstein, 1988).

The authors gratefully acknowledge the role of I. B. Weinstein in the studies with polyoma virus. They also thank Robin Castrogiovanni and Ilse Hoffmann for their valuable assistance in the preparation of this review. This work was supported by grants CA 17613 and CA 51995 from the National Cancer Institute, and SIG8A from the American Cancer Society.
Dedicated to the memory of Michael E. Lambert.

RESPONSE OF DNA TUMOR VIRUSES TO CARCINOGENS

The phenomenon of asynchronous viral DNA replication (AVR), in response to either chemical or physical carcinogenic insult, has been demonstrated for several DNA tumor viruses as summarized in Table 1. AVR is characterized by extensive replication of the viral DNA, which is asynchronous to the cell cycle. This response is also referred to as selective gene amplification due to the increased amount of DNA that is produced in a single cell cycle. In addition, various studies have observed that carcinogen administration leads to DNA replication and transcription of early and late genes, which enable a productive (i.e., lytic) cycle (also referred to as viral reactivation; see Table 1). AVR is observed as an early response that occurs at 12–48 h following DNA damage. This response follows earlier changes in gene expression, which are involved in the regulation of this phenomenon (Ronai et al., 1990). The AVR is observed when two criteria are met: (1) the viral DNA is capable of producing proteins that are essential for its own replication (e.g., large T antigen in the case of polyoma and SV40), and (2) the infected cell is able to provide protein factors that enable replication of the specific viral DNA sequence. Although the first criterion is fully dependent upon the transcription of the viral sequences, cellular protein factors may form complexes with the viral proteins and thereby modulate replication (described in detail below). As demonstrated in this review, it appears that cellular proteins through direct or indirect activities are the limiting factor in viral DNA replication.

The phenomenon of AVR is dose-dependent as increasing concentrations of carcinogens are capable of inducing higher AVR response. This correlation is limited to relative low concentrations, because high doses of carcinogens, which create lesions within the endogenous viral DNA sequences, appear to block the replication of the viral DNA (Berger and Edenberg, 1986; Gough and Wood, 1989; Griffiths and Ling, 1989). This inhibition has been shown to occur by two mechanisms: (1) a blockage of chain elongation by DNA adducts at the replication fork (Mezzina et al., 1988), and (2) an inhibition of the initiation of DNA synthesis in replicons (Kaufmann et al., 1980). Irradiation of SV40-infected cells with UV light (20 J/m^2) induces one to three pyrimidine dimers per SV40 genome, and thus causes a significant inhibition of viral DNA replication. However, when cells are irradiated before virus infection, enhanced replication of the viral DNA results (Treger et al., 1988). Under certain conditions, replication over a damaged site is thought to be the principal source of DNA damage-induced mutagenesis in mammalian cells (Hsia et al., 1989). The dose of carcinogen that is required to create pyrimidine dimers in a 5000-bp DNA sequence (the size of DNA tumor viruses used) results in extensive damage and high cell mortality in most cell systems.

TABLE 1. Responses of DNA Tumor Viruses to Carcinogen Administration

DNA tumor virus	Cell system	Carcinogen	Response	Reference
SV40	CHO	Benzo[a]pyrene	Amplification	Taylor et al. (1982) Bowden et al. (1986)
SV40	CV-1P	UV, benzo[a]pyrene	Reactivation	Taylor et al. (1981)
SV40	CHO	Alpha particle UV radiation	Amplification	Lücke-Huhle et al. (1987)
Polyoma	LPT	UV Mitomycin C	Replication	Fogel and Sachs (1970)
Polyoma	H3	Benzo[a]pyrene	Replication	Lambert et al. (1983)
Polyoma	H3	Nitropyrenes	Replication	Lambert and Weinstein (1987)
Polyoma	H3	UV	Replication	Ronai et al. (1987)
Adeno-associated virus-2 (AAv-2)	CHO K1,XP29	MNNG[a] NAAAF[b] 4-Nitroquinoline N-oxide UV	Amplification	Yalkinoglu et al. (1988)
Adeno-associated	CHO	Chemical carcinogens Cycloheximide	Amplification	Heilbronn et al. (1985)
Adeno-associated	CHO	UV	Replication	Yakobson et al. (1989)
Herpes simplex I	Vero	UV, heat shock	Reactivation	Williams et al. (1989)

[a]N-Methyl-N'-nitro-N-nitrosoguanidine.
[b]N-Acetoxy-2-acetylaminofluorene.

This review will focus on the experiments in which low doses of carcinogens have been applied and in which cell survival is evident.

Because a variety of carcinogens, which are known to produce different types of lesions in the DNA, are capable of inducing AVR (see Table 1), it is likely that the viral response may not be limited to one specific type of DNA damage. Furthermore, given the doses of the carcinogens administered in these studies (see Table 1), it is not likely that the viral DNA sequences are directly affected (because these doses produce DNA lesions at a far lower frequency than the overall size of the DNA tumor viruses studied) (Foils et al., 1992). Yet it is possible that sequences of DNA tumor viruses that are integrated into the cellular genome serve as "hot spots" and are therefore more susceptible to DNA damage.

CARCINOGEN-INDUCIBLE PROTEINS AND VIRAL DNA REPLICATION

The correlation between carcinogen doses, DNA lesions, and the inducibility of AVR as summarized above has given rise to the hypothesis that asynchronous viral DNA replication may be mediated by protein factors that primarily play a role in the regulation of cellular DNA replication.

The effects of carcinogen action on protein expression can be divided into three major time frames: immediate (0–12 h), early (12–48 h), and late (beyond 48 h; see Ronai et al., 1990). The immediate response is characterized by an increased expression of cellular genes, including those involved in DNA repair (reviewed by Bohr, 1987; Hanawalt, 1989; Wallace, 1988), and proto-oncogenes including c-*fos* (Ronai et al., 1988; Bauscher et al., 1988; Hollander and Fornace, 1989), c-*myc* (Ronai et al., 1988), and p-53 (Maltzman and Cztzyk, 1984). Other carcinogen-inducible genes include methallothionine I and II (Fornace et al., 1988), beta-polymerase (Fornace et al., 1989), various heat shock proteins (Brunet and Giacamoni, 1989), and small proline-rich proteins (Gibbs et al., 1990). Numerous studies have demonstrated the appearance of unidentified polypeptides of different sizes that were inducible upon X-ray treatment (Boothman et al., 1989; Lambert and Borek, 1988) or UV irradiation (Schorpp et al., 1984; Herrlich et al., 1986; Lambert et al., 1989). Some of the carcinogen-inducible polypeptides have been identified as secreted factors that are capable of exerting different effects on the expression of certain genes (Schorpp et al., 1984; Rotem et al., 1987; Ronai et al., 1990). We have recently identified a subset of UV-inducible proteins that bind specifically to the polyoma virus regulatory region (Ronai and Weinstein, 1990). Other DNA sequences that play a role in the response of DNA tumor viruses to carcinogens have been recently identified as well (Lücke-Huhle et al., 1989).

Several independent studies have been performed that point to the role of carcinogen-inducible proteins in the replication of DNA tumor viruses. These studies, summarized in Table 2, provide both direct and indirect evidence for the role of proteins in AVR. Initially, various studies provided indirect evidence for the role of proteins in the replication of DNA tumor viruses. In some of these studies, AVR was blocked through administering inhibitors of protein synthesis to carcinogen-treated cells, suggesting that de novo synthesis of cellular proteins is required (Manor and Neer, 1975; Schlehofer et al., 1986; Ronai et al., 1987). Interestingly, administration of low doses of inhibitors, which could block protein synthesis only partially, induced asynchronous replication of viral DNA sequences (Ronai et al., 1987; Yakobson et al., 1989). Because partial inhibition of protein synthesis is likely to primarily affect proteins with a high turnover, the ability to induce AVR through partial blockage of protein synthesis implied that the regulation of viral DNA replication following DNA damage may be mediated by proteins with high turnover. Alternatively, such low concentrations could also stabilize a subset of mRNA transcripts that may affect AVR. More direct support for the role of carcinogen-inducible proteins in the replication of viral DNA sequences comes from experiments in which carcinogen-treated cells were fused with untreated cells that carry an integrated viral DNA sequence. In all reported cases, this has led to asynchronous replication of the viral DNA, suggesting the role of a carcinogen-inducible transacting factor in the replication process (see Table 2).

In related experiments, we introduced protein factors from UV-treated rat fibroblast cells via red-blood-cell-mediated fusion methods into a rat fibroblast cell line (H3) that carries an integrated copy of the polyoma virus. This approach indicated that UV-inducible proteins were capable of inducing, in trans, the replication of the polyoma virus and identified a 60-kD protein fraction that contained this activity (Ronai et al., 1988).

These observations indicate that carcinogen-inducible proteins appear to play a major role in the regulation of asynchronous viral DNA replication. There are three possible mechanisms that could account for this effect. The proteins may interact (1) directly with regulatory sequences within domains positioned on the viral DNA, (2) indirectly with repressor proteins that block viral replication under normal conditions, or (3) through both mechanisms.

To investigate which of these mechanisms predominates, we have utilized methods to measure DNA-protein interactions in which UV-inducible proteins were bound to the regulatory region of the polyoma virus. To this end, we have utilized a 257-bp DNA fragment that contained the enhancer and promoter domains of the polyoma virus as well as part of the origin of replication. When this DNA fragment was incubated with protein extracts prepared 1, 6, 24, and 72 h after UV irradiation of rat fibroblast cells and with control proteins as well, several DNA-

TABLE 2. Evidence for the Role of Stress-Inducible Proteins in the Replication of DNA Tumor Viruses

Treatment	Experimental design	Viral DNA used	Reference
Cycloheximide	Exposure of LPT cells to cycloheximide	Polyoma	Manor and Neer (1975)
UV	Fusion of SV40-transformed BRWSV rat cells with UV-irradiated human diploid fibroblasts	SV40	van der Lubbe et al. (1985)
Benzo[a]pyrene	Fusion of BP-treated rat 6 cells with H3 cells	Polyoma	Lambert et al. (1986a, 1986b)
Cycloheximide, actinomycin D	Exposure of rat H3 cells to low doses of cycloheximide or actinomycin D	Polyoma	Ronai et al. (1987)
UV	Fusion of UV-treated rat 6 fibroblasts with H3 cells	Polyoma	Ronai et al. (1987)
MNNG[a]	Fusion of carcinogen-treated CHO cells with CO60 cells (containing integrated viral DNA)	SV40	Berko-Flint et al. (1988)
UV	Introduction of UV-induced proteins from rat 6 cells, via red-blood-cell-mediated fusion, into H3 cells	Polyoma	Ronai and Weinstein (1988)
UV	UV-induced early domain binding factor as the limiting component of SV40 DNA amplification	SV40	Lücke-Huhle et al. (1989)
UV	Fusion of UV-treated African green monkey cells with SV40-infected cells	SV40	Nomura and Oishi (1989)
MNNG[a]	Cytosolic extracts from MNNG-treated CO60 814 cells supported de novo DNA synthesis in an in vitro replication system	SV40	Berko-Flint et al. (1990)
UV	Identification of UV-inducible proteins that bind to polyoma regulatory region	Polyoma	Ronai and Weinstein (1990)

[a]*N*-Methyl-*N'*-nitro-*N*-nitrosoguanidine.

protein complexes could be identified in gel retardation assays (Ronai and Weinstein, 1990). More striking, however, was the observation that a novel DNA-protein complex was observed with protein extracts prepared 6 and 24 h after UV irradiation. This suggested that a transiently induced/ activated protein(s) binds specifically to the regulatory region of this virus. As it was of interest to identify the exact DNA sequence that binds the UV-inducible protein, a set of DNase I footprinting assays was performed. This led to the identification of two DNA sequences that were protected by factors prepared 6 and 24 h after UV treatment. One of the protected sequences resembled a CAAT motif, which has been shown to play a role in both transcription and DNA replication (Santoro et al., 1988). The second DNA sequence was designated as a UV-response element (URE) with a sequence of TGACAACA. This sequence has common features with the AP1 recognition sequence (GTGAGTa/cA). Nevertheless, examination by means of oligonucleotide competition experiments and immunoprecipitation did not identify the c-*jun* protein among the URE-bound proteins (Ronai and Weinstein, 1990). The UV-induced proteins were purified through affinity chromatography utilizing the URE sequence cross-linked to Sepharose beads. This approach led to the identification of 4 proteins with molecular masses of 40, 55, 62, and 67 kD, with the 40-kD protein as the most abundant. The 62-kD peptide was identified as the c-*fos* protein through immunoprecipitation assays. This finding suggests that c-*fos* may form a heterodimer with another UV-inducible URE-bound protein. The formation of such a complex may increase the affinity of the UV-responsive protein for specific DNA sequences, as has been demonstrated for c-*jun*-c-*fos* (Chiu et al., 1988). The URE is also very similar to the cAMP-responsive element (CRE); however, it is not likely that CRE-bound protein (CREB) is among the URE-bound proteins, because the CREB does not form complexes with c-*fos*, nor does it bind to polyoma DNA sequences (M. Montminy, personal communications). Current studies are aimed at the identification and characterization of the genes encoding the URE-bound proteins through screening a cDNA expression library prepared from UV-treated rat fibroblast cells employing a synthetic URE oligonucleotide probe.

In related studies, Herrlich et al. (1989) have found that UV radiation and other carcinogenic agents induced an increase in DNA binding activity to the early domain of the SV40 minimal origin in both SV40-permissive and SV40-nonpermissive cells. The increase was found to be due to posttranslational modification of a preexisting protein, because it occurs in the presence of cycloheximide and anisomycin (Lücke-Huhle et al., 1989). This study provides evidence that such binding activity may play a role in UV-induced SV40 DNA amplification, a finding that is in agreement with our observation on polyoma DNA replication following UV treatment (Ronai et al., 1990). Additional support for the role of proteins in the response of viral DNA sequences to carcinogens comes from the use of an in vitro replication system that enables the identification of each of the critical components required for AVR. Utilizing such a system, seven cellular proteins that are essential components for the replication of SV40 DNA in vitro have been identified. These include DNA polymerases alpha and delta, topoisomerases I and II, replication factors A and C (RF-A and RF-C), and the proliferation cell nuclear antigen PCNA (reviewed recently by Stillman, 1989). Independent studies have observed increases in expression of DNA polymerases and PCNA following carcinogen administration (Ronai et al., 1990; Tsurimoto and Stillman, 1989). Experiments that employed an in vitro system revealed that cell extracts of MNNG-treated CO60 cells were able to induce higher levels of SV40 replication while extracts from control cells supported only marginal levels of replication (Berko-Flint et al., 1990). Future studies will clarify the nature of these inducible factors and may indicate their mechanism of action on cellular genes that are thought to be their primary target in cells not infected with viral DNA sequences.

An additional carcinogen-responsive element was identified by Elledge and Davis (1989) in the ribonucleotide reductase 2 (RNR2) gene of *Saccharomyces cerevisiae*. This gene is cell cycle-regulated and involved in DNA synthesis after DNA damage. This element has been identified as a 42-bp region that contains both upstream repressor and activator sequences and can confer damage inducibility to a heterogenous promoter.

In addition to changes in DNA synthesis and selective DNA amplification observed following carcinogen administration, recent studies have led to the identification of other carcinogen-responsive elements, positioned on various viral and cellular genes, that are capable of mediating elevated gene expression. Interestingly, UV-responsive elements that were identified on human collagenase, immunodeficiency type 1 (HIV-1), and c-*fos* genes do not share an apparent sequence motif and bind different transacting proteins; a member of the NF_kB family binds to HIV-1 enhancer, the serum response factors p67 and p62 bind to the *fos* promoter, and the heterodimer of *jun* and *fos* binds to the collagenase enhancer (Stein et al., 1989). The role of *jun* and *fos* in the response to UV treatment was also observed in the UV-inducible human small proline-rich II (sprII) gene. The promoter region of the sprII gene was found to contain all the cis elements necessary for induced expression following UV irradiation of cultured human keratinocytes including the AP-1 recognition sequence (Gibbs et al., 1990). Overall, the presence of multiple responsive elements that are activated through divergent transacting proteins provides additional confirmation for the complexity of the response to carcinogens. Nevertheless, the role of these carcinogen-inducible genes with respect to AVR, and perhaps later responses, is yet to be determined.

The identification of the URE in the polyoma system (Ronai and Weinstein, 1990) has led us to search the nucleic acid databases for sequences that are homologous to the URE and that are positioned on the regulatory domain of certain cellular genes. Table 3 shows a list of human cellular genes that carry the URE sequence on their regulatory region. Although it is possible that other, yet unidentified, carcinogen responsive genes are regulated by the URE, elucidating the responses of the genes listed in Table 3 to various carcinogens may shed new light on the nature of carcinogen-inducible responses in mammalian cells.

The observation that carcinogen-inducible proteins are capable of mediating the replication of DNA tumor viruses, together with our identification of a novel DNA binding region on the polyoma DNA, suggests a functional role of the URE sequence in asynchronous replication of polyoma virus. To determine the functional role of the URE following UV irradiation we have synthesized a URE-DNA tetramer that was cloned into a pGEM plasmid. A set of experiments involving cotransfection of a complete polyoma-containing plasmid and the URE-containing plasmid revealed that the URE does play a role in regulating polyoma DNA replication. When the plasmids were cotransfected into normal rat fibroblast cells, the URE was capable of blocking replication of polyoma virus following UV irradiation. Similar studies revealed the role of URE in transcription of polyoma sequences. This result suggests that the UV-inducible protein was out-competed by the synthetic URE sequence. The nature of each of these proteins is currently being studied (Rutberg, et al., 1992). The expression pattern of URE-bound proteins was studied using antibodies that detect primarily the 40 kDa protein. This protein was found to be induced in both rat fibroblasts and human keratinocytes by UV-irradiation as well as heat shock and serum stimulation (Yang et al., 1992).

ASYNCHRONOUS REPLICATION OF DNA TUMOR VIRUSES AS A POSSIBLE INTERMEDIATE STEP FOR STABLE DNA AMPLIFICATION

Although the physical mechanisms involved in asynchronous viral DNA replication have not been clarified, several models to describe this process have been proposed.

Botchan and his colleagues (Miller et al., 1984) proposed the "onion

TABLE 3. Human Genes Carrying the URE Sequence and Their Regulatory Domain

Gene	Position of URE	Homology
Alpha Fetoprotein	180; 5′ flanking region	100%
Alpha amylase	385; 5′ flanking region	100%
Alpha dehydrogenase	466; 5′ flanking	100%

skin" model, in which bidirectional replication at an origin generates a bubble that can undergo further rounds of unscheduled DNA replication. Recombination within this region could generate extrachromosomal circles, which resemble the replication of DNA tumor viruses. A model describing a similar mechanism has recently been proposed by Amler and Schwab (1989), who studied the amplification of the cellular N-myc gene in human neuroblastoma cells. In their study, the amplified N-myc sequences were found to be clustered in tandem repeats, often integrated into various chromosomes at sites other than the normal N-myc gene. Based on these findings, the authors proposed a multistep model to explain the mechanism of amplification of a cellular DNA sequence that resembles the response of DNA tumor viruses to carcinogen administration. The initial step in this model involves an illegitimate replication, recombination and formation of extrachromosomal circular structures. The excised DNA can then replicate extrachromosomally as circular DNA molecules and can thus form double minutes (see also Schimke, 1988; Stark et al., 1989; Wahl, 1989).

Due to similarities between gene amplification and asynchronous viral DNA replication it is tempting to speculate that the transient amplification process, measured by asynchronous replication of DNA tumor viruses, may represent mechanisms involved with the first step in the stable process of gene amplification. This hypothesis is supported, in part, by earlier studies in which cellular genes were selectively amplified following carcinogen administration in a time frame that parallels the asynchronous DNA replication process. For example, Tlsty et al. (1984) observed amplification of the dihydrofolate reductase (DHFR) gene resulting in an increase in the number of methotrexate-resistant colonies in 3T6 murine cells following UV treatment. This effect was transient and reached its maximum 12–24 h after the initial treatment with methotrexate. Similarly, transient amplification of the DHFR gene was observed by Lavi et al. (1987) in Chinese hamster ovary (CHO) cells that were exposed to a chemical carcinogen. It needs to be determined whether the regulatory domains, in cellular genes that are known to amplify in various neoplasms, carry a common regulatory element that responds to some of the carcinogen-inducible proteins as have been identified in the transient experimental processes.

The similarities between the transient asynchronous replication of viral and cellular genes are not limited to their time frame and their karyological changes as they may also share common regulatory pathways. An example for such regulatory proteins is the large T antigen of polyoma and SV40 viruses and the protein product of the tumor suppressor gene p53.

Interestingly, several studies have indicated the interaction of large T antigen with cellular proteins. These interactions can be divided into (1) those involved in the replication complex with DNA polymerase alpha

(Gannon and Lane, 1987) and (2) complexes formed with protein products of the tumor suppressor gene p53 (Tan et al., 1986). The association of large T antigen with p53 is of particular interest as it has been demonstrated to affect the replication of the respective DNA tumor virus (Braithwaite et al., 1987; Sturzbecker et al., 1988; Wang et al., 1989). More recent studies have associated the mutant p53 with stable gene amplification (Wahl et al., 1992). The ability of p53 to interact with p34cdc2 (Sturzbecker et al., 1990) suggests a novel regulatory mechanism in which T antigen may affect signal transduction pathways and cell cycle control.

While large T antigen have also been associated with amplification of cellular genes (Miller et al., 1984; Pellegrini and Basilico, 1987; Robinson and Ronai, 1990), the mechanisms involved in this process have not been elucidated as yet. The recent observations of protein-protein complexes composed of viral-cellular proteins points to a novel mechanism involved in the regulation of AVR and perhaps other cellular responses to carcinogens.

SUMMARY

The complex response of mammalian cells to carcinogens is mediated through DNA replication and selective gene expression. This review demonstrated the role of carcinogen inducible proteins as mediators of viral DNA replication, as a model system to understand changes in cellular genes which carry similar regulatory domains.

It is clear that the response to carcinogens includes multiple genes, which are regulated through divergent pathways. The availability of methodologies to understand protein-DNA interactions has uncovered some of the critical steps in the regulation of gene expression and replication. It has been demonstrated that various cellular changes following carcinogen treatment modulate DNA-protein interaction through newly induced proteins as well as through posttranslational modifications of preexisting regulatory proteins. The model systems and hypotheses described in this review, together with new advances in understanding protein-protein complexes, signal transduction cascades, and cell cycle control, will lead to progress in this important area of mutagenesis and carcinogenesis.

REFERENCES

Amler, L. C., and Schwab, M. 1989. Amplified N-myc in human neuroblastoma cells is often arranged as clustered tandem repeats of recombined DNA. *Mol. Cell. Biol.* 9:4903–4913.

Bauscher, M., Rahmsdorf, H. J., Litfin, M., Karin, M., and Herrlich, P. 1988. Activation of the c-fos gene by UV and phorbol ester; different signal transduction pathways coverage to the same enhancer element. *Oncogene* 3:301–311.

Berger, C. A., and Edenberg, H. J. 1986. Pyrimidine dimers block simian virus 40 replication forks. *Mol. Cell. Biol.* 6:3443–3450.

Berko-Flint, Y., Karby, S., and Lavi, S. 1988. Carcinogen induced factors responsible for SV40 DNA replication and amplification in Chinese hamster cells. *Cancer Cells* 6:183–189.

Berko-Flint, Y., Karby, S., Hassin, D., and Lavi, S. 1990. Carcinogen induced DNA amplification in vitro: Overreplication of the simian virus 40 origin region in extracts from carcinogen-treated CO60 cells. *Mol. Cell. Biol.* 10:75–83.

Bohr, V. A. 1987. Differential DNA repair within the genome. *Cancer Rev.* 7:28–55.

Boothman, D. A., Bouvard, I. D., and Hughes, E. N. 1989. Identification and characterization of x-ray induced proteins in human cells. *Cancer Res.* 49:2871–2878.

Bowden, G. T., Ossanna, N., and Hurd, E. 1986. Benzo(a)pyrene diol epoxide induces viral reactivation at concentrations that block DNA elongation in mammalian cells. *Chem. Biol. Interact.* 58:333–344.

Braithwaite, A. W., Sturzbecker, H. W., Addison, C., Palmer, C., Rudge, K., and Jenkins, J. R. 1987. Mouse P53 inhibits SV40 origin dependent DNA replication. *Nature* 329:458–460.

Brunet, S., and Giacamoni, P. U. 1989. Heat shock mRNA in mouse epidermis after UV-irradiation. *Mutat. Res.* 219:209–216.

Chiu, R., Boyle, W. J., Meek, J., Smeal, T., Hunter, T., and Karin, M. 1988. The c-Fos protein interacts with c-Jun/AP-1 to simulate transcription of AP-1 responsive genes. *Cell* 54:541–552.

Elledge, S. J., and Davis, R. W. 1989. Identification of the DNA damage-responsive element of RNR 2 and evidence that four distinct cellular factors bind it. *Mol. Cell. Biol.* 9:5373–5386.

Fogel, M., and Sachs, L. 1970. Induction of virus synthesis in polyoma transformed cells by ultraviolet light and mitomycin C. *Virology* 38:174–177.

Foiles, P. G., Peterson, L. A., Miglietta, L. M., and Ronai, Z. A. 1992. *Mutat. Res.,* in press.

Fornace, A. J., Jr., Alamo, I., Jr., and Hollander, M. C. 1988. DNA damage-inducible transcripts in mammalian cells. *Proc. Natl. Acad. Sci. USA* 85:8800–8804.

Fornace, A. J., Jr., Zmudzka, B., Hollander, M. C., and Wilson, S. H. 1989. Induction of beta-polymerase mRNA by DNA-damaging agents in Chinese hamster ovary cells. *Mol. Cell. Biol.* 9:851–853.

Gannon, J. V., and Lane, D. P. 1987. p53 and DNA polymerase alpha compete for binding to SV40 T antigen. *Nature* 329:456–458.

Gibbs, S., Lohman, F., Teubel, W., Van de Putte, P., and Backendorf, C. 1990. Characterization of the human spr2 promoter: Induction after UV irradiation or TPA treatment and regulation during differentiation of cultured primary keratinocytes. *Nucleic Acids Res.* 18:4401–4407.

Gough, G., and Wood, R. D. 1989. Inhibition of in vitro SV40 DNA replication by ultraviolet light. *Mutat. Res.* 227:193–197.

Griffiths, T. D., and Ling, S. Y. 1989. Effect of UV light on DNA chain growth and replication initiation in human cells. *Mutat. Res.* 218:87–94.

Hanawalt, P. C. 1989. Concepts and models for DNA repair from *Escherichia coli* to mammalian cells. *Environ. Mol. Mutagen.* 14:90–98.

Heilbronn, R., Schlehofer, J. R., Yalkinoglu, A. O., and Zur-Hausen, H. 1985. Selective DNA-amplification induced by carcinogens (initiators): Evidence for a role of proteases and DNA polymerase alpha. *Int. J. Cancer* 36:85–91.

Herrlich, P., Angel, P., Rahmsdorf, H. J., Mallick, U., Poting, A., Heiber, L., Lücke-Huhle, C., and Schorpp, M. 1986. The mammalian genetic stress response. *Adv. Enzyme Regul.* 25:485–495.

Hollander, M. C., and Fornace, A. J., Jr. 1989. Induction of fos RNA by DNA-damaging agents. *Cancer Res.* 49:1687–1692.

Hsia, H. C., Lebkowski, J. S., Leong, P. M., Calos, M. P., and Miller, J. H. 1989. Comparison of ultraviolet irradiation induced mutagenesis at the LacZ gene in *Escherichia coli* and in human 293 cell. *J. Mol. Biol.* 205:102–113.

Kaufmann, W. K., Cleaver, J. E., and Painter, R. B. 1980. UV inhibits replicon initiation in S phase human cells. *Biochim. Biophys. Acta* 608:191–195.

Lambert, M., and Borek, C. 1988. X-ray induced changes in gene expression in normal and oncogene-transformed rat cells lines. *JNCI* 80:1492–1497.

Lambert, M. E., Garrels, J. I., McDonald, J., and Weinstein, I. B. 1986a. Inducible cellular responses to DNA damage in mammalian cells. In *Antimutagenesis and Anti-Carcinogenesis Mechanisms*, eds. D. N. Shankel and P. E. Hartman, pp. 291–311. New York: Plenum Press.

Lambert, M. E., Gattoni-Celli, S., Kirschmeier, P., and Weinstein, I. B. 1983. Benze[a]pyrene induction of extrachromosomal viral DNA synthesis in rat cells transformed by polyoma virus. *Carcinogenesis* 4:587–593.

Lambert, M. E., Pellegrini, S., and Weinstein, I. B. 1986b. Carcinogen induced asynchronous replication of polyoma DNA is mediated by a trans acting factor. *Carcinogens* 7:1011–1017.

Lambert, M. E., Ronai, Z. A., Weinstein, I. B., and Garrels, J. I. 1989. Enhancement of major histocompatibility Class I protein synthesis by DNA damage in cultured human fibroblasts and keratinocytes. *Mol. Cell. Biol.* 9:847–850.

Lambert, M. E., and Weinstein, I. B. 1987. Nitropyrenes are inducers of polyoma viral DNA synthesis. *Mutat. Res.* 183:203–311.

Lavi, S., Kleinberger, T., Berko-Flint, Y., and Blamte, M. 1987. Stable and transient amplification of dhfr and SV40 in carcinogen treated SV40 transformed cells. In *Accomplishments in Oncology*, vol. 2, eds. H. Zur Hausen and J. R. Schlehofer, pp. 117–125. Philadelphia: J. B. Lippincott.

Lücke-Huhle, C., and Herrlich, P. 1987. Alpha-radiation induced amplification of integrated SV40 sequences is mediated by a trans-acting mechanism. *Int. J. Cancer* 39:94–98.

Lücke-Huhle, C., Mai, S., and Herrlich, P. 1989. UV-induced early domain binding factor as the limiting component of simian virus 40 DNA amplification in rodent cells. *Mol. Cell. Biol.* 9:4812–4818.

Maltzman, W., and Cztzyk, L. 1984. UV-irradiation stimulates levels of p53 cellular tumor antigen in nontransformed mouse cells. *Mol. Cell. Biol.* 4:1689–1694.

Manor, H., and Neer, A. 1975. Effects of cycloheximide on virus DNA replication in an inducible line of polyoma transformed rat cells. *Cell* 5:311–318.

Mezzina, M., Menck, C. F., Courtin, P., and Sarasin, A. 1988. Replication of simian virus 40 DNA after UV irradiation: Evidence of growing fork blockage at single stranded gaps in daughter strands. *J. Virol.* 62:4249–4258.

Miller, J., Bullock, P., and Botchan, M. 1984. Simian virus 40 T antigen is required for viral excision from chromosomes. *Proc. Natl. Acad. Sci. USA* 81:7534–7538.

Nomura, S., and Oishi, M. 1984. UV-irradiation induces an activity which stimulates simian virus 40 rescue upon cell fusion. *Mol. Cell. Biol.* 4:1159–1162.

Pellegrini, S., and Basilico, C. 1987. Amplification and expression of foreign genes in cells producing polyoma virus large T antigen. *Oncogene Res.* 1:23–41.

Pellegrini, S., Dailey, L., and Basilico, C. 1984. Amplification and excision of integrated polyoma DNA sequences require a functional origin of replication. *Cell* 36:943–949.

Robinson, R., and Ronai, Z. 1990. Replication of polyoma DNA sequences is limited to ascites form of SEWA sarcoma cells. *Mol. Carcinogen.*, in press.

Ronai, Z. A., Lambert, M. E., Johnson, M. D., Okin, E., and Weinstein, I. B. 1987. Induction of asynchronous replication of polyoma DNA in rat cells by ultraviolet light irradiation and the effect of various inhibitors. *Cancer Res.* 47:4565–4570.

Ronai, Z. A., Lambert, M. E., and Weinstein, I. B. 1990. Inducible cellular responses to ultraviolet light irradiation and other mediators of DNA damage in mammalian cells. *Cell. Biol. Toxicol.* 6:105–126.

Ronai, Z. A., Okin, E., and Weinstein, I. B. 1988. Ultraviolet light induces expression of oncogenes in rat fibroblasts of human keratinocytes cells. *Oncogene* 2:201–204.

Ronai, Z. A., and Weinstein, I. B. 1988. Identification of a UV-induced trans-acting protein that stimulates polyoma virus DNA replication. *J. Virol.* 62:1057–1060.

Ronai, Z. A., and Weinstein, I. B. 1990. Identification of ultraviolet inducible proteins that bind to a TGACAACA sequence in the polyoma virus regulatory region. *Cancer Res.* 50:5374–5381.

Rotem, N., Axelrod, J. H., and Miskin, R. 1987. Induction of urokinase type plasminogen activator by UV light in human fetal fibroblasts is mediated through a UV-induced secreted protein. *Mol. Cell. Biol.* 7:622–631.

Rutberg, S. E., Yang, Y. M., and Ronai, Z. 1992. Functional role of the URE (TGACAACA) in replication and transcription of polyoma DNA sequences. *Nucleic Acids Res.,* in press.

Santoro, C., Mermond, N., Andrews, P. C., and Tjian, R. 1988. A family of human CCAAT-box binding proteins active in transcription and DNA replication:cloning and expression of multiple cDNAs. *Nature* 334:218–224.

Schimke, R. T. 1988. Gene amplification in cultured cells. *J. Biol. Chem.* 263:5989–5992.

Schlehofer, J. R., Ehrbar, M., and Zur Hausen, H. 1986. Vaccinia virus, herpes simplex virus and carcinogens induce DNA amplification in a human cell line support replication of a helper virus dependent parvo virus. *Virology* 151:110–117.

Schorpp, M., Mallick, U., Rahmsdorf, H. J., and Herrlich, P. 1984. UV induced extracellular factor from human fibroblasts communicates the UV response to non-irradiated cells. *Cell* 37:861–868.

Stark, G. R., Debatisse, M., Giulotto, E., and Whal, G. M. 1989. Recent progress in understanding mechanisms of mammalian DNA amplification. *Cell* 57:901–908.

Stein, B., Kramer, M., Rahmsdorf, H. J., Ponta, H., and Herrlich, P. 1989. UV-induced transcription from the human immunodeficiency virus type-1 (HIV-1) long terminal repeat and induced secretion of an extracellular factor that induces (HIV-1) transcription in nonirradiated cells. *J. Virol.* 63:4540–4544.

Stillman, B. 1989. Initiation of eukaryotic DNA replication *in vitro. Annu. Rev. Cell. Biol.* 5:197–245.

Sturzbecker, H. W., Brain, R., Maimets, T., Addison, C., Rudge, K., and Jenkins, J. R. 1988. Mouse P53 blocks SV40 DNA replication *in vitro* and down regulates T antigen DNA helicase activity. *Oncogene* 3:405–413.

Sturzbecker, H. W., Maimets, T., Chumaker, P., Brain, R., Addison, C., Simans, V., Rudge, K., Philip, R., Grimaldi, M., Court, W., and Jenkins, J. K. 1990. p53 Interacts with p34 cdc2 in mammalian cells; Implications for cell cycle control and oncogenesis. *Oncogene* 5:795–810.

Taylor, W. E., and Sanford, K. K. 1981. Effect of visible light on benzo(a)pyrene binding to DNA of cultured human skin epithelial cells. *Cancer Res.* 41:1789–1793.

Tlsty, T. D., Brown, P. C., and Schimke, R. T. 1984. UV-radiation facilitates methotrexate resistance and amplification of the dihydrofolate reductase gene in cultured 3T6 mouse cells. *Mol. Cell. Biol.* 4:1050–1056.

Treger, J. M., Hauser, J. V., and Dixon, K. 1988. Molecular analysis of enhanced replication of UV-damaged simian virus 40 DNA in UV-treated mammalian cells. *Mol. Cell. Biol.* 8:2428–2434.

Tsurimoto, T., and Stillman, B. 1989. Multiple replication factors augment DNA synthesis by the two eukaryotic DNA polymerases, alpha and delta. *EMBO J.* 8:3883–3889.

Van der Lubbe, J. L. M., Van Drunen, C. M., Herloghs, J. J. L., Cornelis, J. J., Rommelaere, J., and Van der Eb, A. J. 1985. Enhanced induction of SV40 replication from transformed mammalian cells by fusion with UV-irradiated untransformed cells. *Mutat. Res.* 151:1–8.

Wahl, G. N. 1989. The importance of circular DNA in mammalian gene amplification. *Cancer Res.* 49:1333–1340.

Wahl, G. M., Yin, Y., Tainsky, M. A., Bisohoff, F. Z., McGill, J., Forseth, B., and Van Hoff, D. D. 1992. *Proceedings of the American Association for Cancer Research.* 33:584–585.

Wallace, S. S. 1988. AP endonucleases and DNA glycosylases that recognize oxidative DNA damage. *Environ. Mol. Mutagen.* 12:431–477.

Wang, E. H., Friedman, P. N., and Prives, C. 1989. The murine p53 protein blocks replication of SV40 DNA in vitro by inhibiting the initiation functions of SV40 large T antigen. *Cell* 57:379–392.

Weinstein, I. B. 1988. The origins of human cancer: Molecular mechanisms of carcinogenesis and their implications for cancer prevention and treatment, twenty-seventh G.H.A. Clows Memorial Award Lecture. *Cancer Res.* 48:4135–4143.

Williams, J. J., Landgraf, B. E., Whiting, N. L., and Zurlo, J. 1989. Correlation between the induction of heat shock protein 70 and enhanced viral reactivation in mammalian cells treated with ultraviolet light and heat shock. *Cancer Res.* 49:2735–2742.

Yakobson, B., Hrynko, T. A., Peak, J. J., and Winocour, E. 1989. Replication of adeno-associated virus cells irradiated with UV light at 254 nm. *J. Virol.* 63:1023–1030.

Yalkinoglu, A. O., Heilbronn, R., Burkle, A., Schlehofer, J. R., and Zur Hausen, H. 1988. DNA amplification of adeno-associated virus as a response to cellular genotoxic stress. *Cancer Res.* 48:3123–3129.

Yang, Y. M., Rutberg, S. E., Foiles, P. G., and Ronai, Z. 1992. Expression pattern of proteins that bind to the URE in human keratinocytes. *Mol. Carcinogen,* in press.

8 CARCINOGEN-INDUCED GENE AMPLIFICATION AND GENE EXPRESSION

Mirit I. Aladjem, Sara Lavi

Department of Microbiology/Cell Biology, The George S. Wise Faculty of Life Sciences, Tel Aviv University, Ramat Aviv, Israel

INTRODUCTION

The genome of living organisms is subject to a continuous threat from the environment. Chemical and physical carcinogens, toxic compounds, and infectious agents may cause significant alterations in the chemical structure and organization of the genetic material. Oncogenic transformation may be caused by carcinogen-induced mutations, amplifications and rearrangements of specific genes (Balmain and Brown, 1988; Cooper, 1990), or by the action of specific tumor viruses (Benjamin and Vogt, 1990). To counter the deleterious effects of environmental carcinogenesis, living organisms have evolved an array of damage-control functions. These range from chemical inactivation of the genotoxic compounds and the repair of carcinogen-induced lesions to the control of viral infections. The higher incidence of oncogenic transformation in patients suffering from various syndromes that impair DNA repair mechanisms (Defais, 1990) or affect the ability to control viral infection (Orth et al., 1979) demonstrates the crucial importance of such defense mechanisms in normal subjects. To gain insight into the molecular mechanisms leading to carcinogen-induced transformation and the role of defense processes in the prevention of such changes, our group has been studying the biochemical changes that follow the exposure to carcinogens at the cellular level. Transient changes in genome stability and altered patterns of gene expression in carcinogen-treated cells suggest that the exposure to carcinogens leads to major perturbations in cellular control processes.

This research was supported by a grant from the Israel Academy of Sciences and from the Israel Association for Cancer Research. S. Lavi is a recipient of a career development award from the Israel Cancer Research Fund.

Present address, to whom reprint requests should be addressed: M. I. Aladjem is Department of Chemical Immunology, The Weizmann Institute of Science, Rehovot, Israel.

CARCINOGEN-INDUCED GENE AMPLIFICATION

Genomic instability, manifested as chromosomal breaks, translocations, and amplifications, is abundant in tumor cells. In some cases, translocation and amplification of specific oncogenes in chemically induced tumors are implicated in the transformation process (Brown et al., 1990; Alitalo and Schwab, 1985). To study the direct induction of gene amplification by exposure to carcinogenic agents, several systems were developed in our laboratory utilizing both viral and cellular sequences as markers.

Amplification of Viral Genes

The SV40 genome, integrated into Chinese hamster embryo cells (in the CO60 cell line), was used as a marker to follow the response of DNA sequences to carcinogen treatment. In these cells, which are semipermissive for SV40 infection, the viral DNA is not capable of autonomous replication and is duplicated as a part of the cellular genome during S phase. An increase in the copy number of SV40 sequences was observed in this system using DNA hybridization (Lavi, 1981; Lavi and Etkin, 1981), thus establishing a direct link between exposure to carcinogens and the amplification phenomenon. This assay facilitates direct monitoring of the clones amplification process without the need to select for cell cones that contain amplified genes.

SV40 amplification is dose dependent, is induced by various carcinogens that differ in their chemical composition, and requires the presence of an intact SV40 origin of replication in cis and T antigen in trans. The amplification is a transient phenomenon, which can be detected by 36 h posttreatment, peaks at 72 h, and gradually diminishes thereafter.

Carcinogen-induced overreplication is not unique to SV40. An analogous system using polyoma virus as a genetic marker in semipermissive rat cells shows similar overreplication of the integrated polyoma genome following carcinogen treatment (Ronai et al., 1987). Adenoassociated virus, which usually requires a helper in its replication, can forego this requirement when infecting carcinogen-treated cells (Schlehofer et al., 1986; Yakobson et al., 1989).

The carcinogenic compounds that were tested for the ability to induce amplification vary in their chemical structure and the nature of their interaction with DNA. These include chemical carcinogens and chemotherapeutic agents that directly interact with nucleic acids and chemicals that inhibit cell growth by indirectly affecting DNA synthesis. In addition, viral agents, such as herpes simplex virus and vaccinia virus, can also induce SV40 amplification in Chinese hamster cells (Schlehofer et al., 1986; Matz et al., 1984). In contrast, tumor promoters that do not have initiating carcinogenic activity, such as phorbol esters, do not induce SV40 amplification. Table 1 lists some examples of various carcinogens that were shown to cause SV40 amplification in our system.

The time scale of amplification and its transient nature suggest that the process is mediated through carcinogen-induced cellular factors. To directly test this hypothesis, a series of cell fusion experiments demonstrated that amplification could be induced in trans by fusing carcinogen-treated cells that do not contain SV40 sequences with SV40-transformed cells that were not exposed to carcinogens (Berko-Flint et al., 1988). In these experiments, a carcinogen with a short half-life was used, thus avoiding possible direct interaction between the carcinogen and the cells harboring viral DNA. Such fusion led to the induction of SV40 overreplication in the nontreated cells, thus demonstrating the existence of carcinogen-induced cellular transactivator(s) of amplification. Furthermore, we showed that Chinese hamster embryo cells, which usually do not support replication of SV40 virus, allow such replication after exposure of the cells to carcinogens prior to infection (Berko-Flint et al., 1988). Thus, carcinogen-induced SV40 overreplication probably stems from an altered cellular, and not viral, activity. The identification of this carcinogen-induced trans-acting activity that is responsible for the amplification of SV40 is one of the main objectives of research in our laboratory. Recent progress in the characterization of this activity utilizing an in vitro amplification system is reported later in this chapter.

Amplification of Cellular Genes

The studies just discussed used SV40 sequences as markers, exploiting the advantages of this system in the rapid detection of amplified sequences shortly after exposure to carcinogens. To investigate whether our findings in the SV40 system could be extended to nonviral markers, the amplification of the *dhfr* gene was followed after exposure to chemical carcinogens. Our studies demonstrate that the magnitude of carcinogen-induced amplification of these and other cellular sequences did not allow the direct detection of altered copy number using crude hybridization techniques. However, amplification of the *dhfr* gene could be detected by measuring the frequency of methotrexate-resistant colonies in carcinogen-treated cell populations. Methotrexate-resistant colonies harboring amplified *dhfr* sequences were detected in both carcinogen-treated and nontreated cells, but the frequency of resistant colonies was increased by up to two orders of magnitude following exposure to carcinogens (Kleinberger et al., 1986). In our studies, we mainly used *N*-methyl-*N'*-nitro-*N*-nitrosoguanidine (MNNG) or UV irradiation as amplification-inducing agents. Studies using mouse 3T3 cells (Tlsty et al., 1984) demonstrated that the DNA replication inhibitor hydroxyurea was also able to increase the frequency of resistant colonies. Further studies

TABLE 1. Carcinogenic Compounds and their Effect on Genome Stability

Class	Examples	Specific effect
Alkylating agents:		
Minor methylating and alkylating	N-methyl-N'-nitro-N-nitrosoguanidine (MNNG)	SV40, *dhfr* and CupI amplification, AAV replication
		Enhanced permissivity to SV40 replication in vivo and in vitro
		Prolonged S-phase and cell cycle arrest
		Enhanced expression of SV40, *dhfr* and transfected plasmids
	N-methylnitrosourea (MNU)	SV40 amplification
	Ethylmethane Sulfonate (EMS)	SV40, *dhfr* and CupI amplification
		Enhanced expression of *dhfr*
		Prolonged S-phase and cell cycle arrest
Crosslinking	*cis*-Diaminochloroplatin II	SV40 amplification
		Enhanced expression of transfected plasmids
Bulky adduct formers:		
Polycyclic hydrocarbons	Benz(a)pyrene diol epoxide	SV40 amplification
		Prolonged S-phase and cell cycle arrest
	7,12-dimethylbenzanthracene (DMBA)	SV40 and *dhfr* amplification
		Enhanced expression of *dhfr* and transfected plasmids
		Enhanced permissivity to infecting SV40 replication

Aromatic amines and nitro compounds	
N-acetoxy-acetyl-aminofluorene (AAAF)	SV40 amplification
4-Nitroquinoline oxide	SV40 and CupI amplification
Intercalating agents	
Ethidium bromide (EtBr)	SV40 amplification
Adriamycin	SV40 amplification
	Enhanced expression of transfected plasmids
Thymidine dimer inducer	
UV irradiation	SV40 and *dhfr* amplification
	Increased expression of *dhfr* and transfected plasmids
	Enhanced permissivity to SV40 replication in vitro
Clastogenic agents	
r-irradiation	SV40 amplification
Bleomycin	SV40 amplification
DNA synthesis inhibitors:	
Specific inhibitors	
Hydroxyurea	SV40 and *dhfr* amplification
Aphidicolin	SV40 amplification
Metabolic inhibitors	
Methotrexate	SV40 and *dhfr* amplification
	Enhanced expression of *dhfr*
Cycloheximide	SV40 and *dhfr* amplification
	Enhanced expression of *dhfr*
ADP-ribosylation inhibitor	
3-aminobenzamide	SV40 amplification
Other compounds	
Diethylstilbestrol	SV40 amplification

117

demonstrated that this increase was not limited to genotoxic agents, as other factors, such as hypoxia, were shown to induce a similar phenomenon (Johnston et al., 1986). SV40 and *dhfr* amplification shared common dose-response and time-course behavior (Kleinberger et al., 1986), suggesting that common cellular factors are involved in the induction of both phenomena.

Carcinogen-induced amplification was also detected in a lower eukaryote, the yeast *Saccharomyces cerevisiae*. In this organism the *Cupl* gene was used as a genetic marker, as the phenotype of yeast colonies containing amplified *Cupl* copies is readily detected by copper resistance. Exposure of a logarithmically growing yeast population to MNNG, ethylmethanesulfonate (EMS), or 4-nitroquinoline 1-oxide (4NQO) increased the frequency of copper-resistant colonies by two orders of magnitude (Aladjem et al., 1988).

The physical structure of the amplified sequences in both carcinogen-treated and untreated copper-resistant colonies is similar. Thus it seems that exposure to carcinogens increases the frequency of a rare but extant cellular process, rather than inducing a novel pathway. A similar conclusion can be drawn concerning *dhfr* amplification in Chinese hamster cells. Amplification of *dhfr* is detected in methotrexate-resistant cells from carcinogen-treated and untreated populations, though at a higher frequency in the former. In the SV40 system, the requirement for an intact origin of replication and an active T antigen for amplification is similar to that observed in normal SV40 replication in permissive cells. Thus, it is plausible that this system also responds to a cellular trigger by activating an existing pathway of viral development. This conclusion is further supported by analysis of SV40 gene expression, as described in the next section.

CARCINOGEN-INDUCED ALTERATIONS IN GENE EXPRESSION AND CELL CYCLE

Our studies utilizing SV40-transformed Chinese hamster cells demonstrate that exposure to chemical carcinogens alters the expression pattern of the integrated viral copy. Analyses of both nascent transcripts and steady-state mRNA indicate that there is a constitutive level of expression of the SV40 early genes in SV40-transformed cells, while the late viral genes are not detectably transcribed. Carcinogen treatment considerably enhances the transcription from the early coding region and induces the transcription of the late genes (Aladjem and Lavi, in press). Overtranscription and overreplication share cis requirements (the viral origin of replication) and are sensitive to DNA replication inhibition by aphidicolin. This interdependence of replication and transcription may point toward the involvement of a common pathway in their activation by a carcinogen-induced cellular factor.

Specific characteristics of carcinogen-induced gene expression are shared with the lytic behavior of the SV40 genome in a permissive environment. These include the coregulation of transcription and replication, the induction of late mRNA synthesis, and the use of RNA polymerase II as the sole enzyme catalyzing overtranscription. Thus, carcinogen-induced SV40 overreplication and overtranscription could be viewed as the result of an overall relaxation of the cellular mechanisms that are responsible for the nonpermissivity of Chinese hamster cells to the development of this virus. This increased permissivity was also manifested in our in vitro studies described in the next section.

To determine whether carcinogen-induced overexpression is extended to nonviral sequences, we measured the expression of selected transfected and endogeneous genetic markers following carcinogen treatment. The expression of transfected plasmids harboring the chloramphenicol acetyltransferase (CAT) gene linked to the SV40 promoter, as well as to the cellular β-actin promoter or to the LTR of an intracisternal A particle, was measured in carcinogen-treated cells (Kleinberger et al., 1988a). Our results indicate that exposure to the carcinogen prior to transfection resulted in enhanced expression of the CAT marker. The enhanced expression peaks at 72 h posttreatment, thus showing a kinetic behavior similar to that observed for SV40 overreplication (discussed in the preceding section). Hence, trans-acting cellular factors are involved in the stimulation of both overexpression and overreplication.

The expression of the cellular dhfr gene was measured using fluorescent methotrexate. We found that the proportion of cells containing an enhanced dhfr level increases significantly after exposure to carcinogens (Kleinberger et al., 1988b). This finding provided us with the ability to sort out a population of cells in which the dhfr content was enhanced. Using this approach, cells were separated according to their methotrexate binding, and the copy number of SV40 sequences was measured by hybridization. It was found that the copy number of SV40 sequences was increased in the cell population that showed enhanced level of fluorescent methotrexate binding. Thus, the same population of carcinogen-exposed cells, in which amplification-mediating cellular factor(s) are induced, may be prone to both SV40 amplification and dhfr overexpression. Similarly, cell sorting experiments indicate that the expression of an integrated CAT, linked to an SV40 early promoter, is elevated in the same population of cells in which the dhfr content increases after exposure to carcinogens, and that this population is more prone to SV40 amplification (Kleinberger et al., 1988b). Thus, the carcinogen-induced expression of these two unlinked genes is probably triggered by the same trans activator.

The carcinogen-enhanced expression of transfected sequences was demonstrated in SV40-transformed Chinese hamster cells, in the stable cell line CHO and in primary embryonic Chinese hamster cultures (Kleinberger et al., 1988a; Melamed, 1988). Thus, enhanced expression, although correlated with increased permissivity to SV40 development, is not limited to genes that are activated by viral promoters, or to cells that are transformed by DNA tumor viruses. Studies by other groups also demonstrated the enhanced expression of viral and cellular genes following exposure to chemical and physical carcinogens (Valerie et al., 1988; Herrlich et al., 1986).

Major perturbations in overall cellular reactions are observed in carcinogen-treated cells. Following exposure to carcinogens, a prolonged S phase is observed, and a slowdown in the progression through the cell cycle ensues (Berko-Flint et al., 1988). Studies in Chinese hamster (CO60) cells exposed to the alkylating agent EMS demonstrated that it takes the EMS-treated cells up to 3 days to advance through a single cell cycle, compared with an 18-h cell cycle in untreated cells. This change in cell cycle length is correlated with SV40 amplification. Synchronized cells that were exposed to carcinogens in G1 or early S phase showed marked elevation in their SV40 content compared with nonsynchronized carcinogen-treated cells or cells treated in the G2 phase (Berko-Flint, 1988). Similarly, the frequency of *dhfr* amplification is enhanced when cells are treated with carcinogens during the S phase (Kleinberger et al., 1986). Thus, carcinogen-induced prolongation of the cell cycle may contribute toward the enhanced level of amplification in general and particularly to the increased permissivity to SV40 observed in carcinogen-treated cells. Further studies in our laboratory are aimed at the identification and characterization of the cellular activity that directly mediates the permissive response at the biochemical level.

TOWARD IDENTIFICATION OF CARCINOGEN-INDUCED TRANSACTIVATORS IN A CELL-FREE SYSTEM

The altered patterns of SV40 replication and gene expression following carcinogen treatment suggest that exposure to carcinogens may cause increased permissivity to SV40 development in a hitherto semipermissive environment. To directly study this hypothesis, and to provide possible means to the identification of the cellular factor(s) that may be involved in cellular permissivity, we developed a cell-free system. Earlier studies by Li and Kelly demonstrated that the ability to support SV40 replication in vitro was directly correlated with cell permissivity to viral replication in vivo (Li and Kelly, 1985). Thus, extracts from primate cells, but not from rodent cells, could support SV40 replication in vitro. We investigated whether extracts from carcinogen-treated Chinese hamster cells could support in vitro replication of SV40 in a manner analogous to extracts from permissive cells. Indeed, an increase in the permissivity to SV40 replication in vitro was observed (Berko-Flint et al., 1990). The replication supported by extracts from carcinogen-treated Chinese hamster cells is limited in the extent of elongation of DNA template compared with that observed using extracts from primate cells; nevertheless, it shares the cis and trans requirement of the permissive system for origin sequences and external T antigen. Extracts from Chinese hamster cells that were not treated with carcinogens are totally nonpermissive to SV40 replication.

The ability to mimic in vitro the increased permissivity observed in vivo provides us with an important tool in our investigation of the carcinogen-induced pleiotropic activity that is capable of amplification and altered gene expression. To this end we initiated a series of experiments aimed at identification of the enzymatic activity which is responsible for the increased permissivity to in vitro replication. Extracts from carcinogen-treated versus nontreated cells do not differ in the rate of replication of artificially primed templates or in their ability to prime single-stranded DNA templates. Conversely, only extracts from carcinogen-treated cells are able to specifically initiate replication from the SV40 origin. An origin-unwinding activity that is present in the extracts from carcinogen-treated cells, but absent in extracts from control cells, is responsible for this change in replication ability (M. I. Aladjem and S. Lavi, unpublished). Thus, the enhanced capability to support replication could be attributed to the ability to present the template molecule in a topological form that will allow further processing by the regular cellular replication machinery.

The induction of an activity that alters the topological structure of the SV40 origin of replication could be responsible for enhanced transcription as well as replication of the virus, by activating the viral early and late promoters that lie adjacent to the origin of replication. Thus, if the in vitro system reflects the intracellular situation, the increased permissivity in vivo could be envisaged as the result of such a topological activation, or the removal of a steric hindrance that prevented the onset of viral ontogeny in the nonpermissive environment. This change in permissivity might reflect an overall stress-induced relaxation of the mechanisms that tightly regulate the replication and transcription in normal situations. Upon exposure to genotoxic compounds, such a relaxation may be needed to allow the action of repair mechanisms and prevent the loss of important genetic information. The changes in replication and gene expression described in our experimental system might all be attributed to the pleiotropic effects of a cellular regulator that is activated upon carcinogen treatment. Further characterization of the in vitro permissivity factor may lead to better insight into the nature and trigger of this cellular regulator.

CONCLUSION

Our data demonstrate that exposure to carcinogens in a variety of cell types results in amplification and overexpression of selected DNA sequences. Our in vivo and in vitro results point toward the involvement of specific cellular regulatory factors in the activation of the integrated SV40 genome in Chinese hamster cells. Kinetic considerations and the results of flow cytometric analyses suggest that the cell population in which these factors are activated is more prone to the amplification and overexpression of cellular genes. Thus, activation of a common cellular trigger may result in several distinct detectable end points, such as gene amplification, overexpression, and enhanced virus permissivity.

The obvious similarities between SV40 induction and bacteriophage lambda induction in bacterial cells also suggest the existence of a pleiotropic activity similar to the bacterial SOS pathway in mammalian cells. Perturbation of the cell cycle, observed following carcinogen treatment, may be triggered by such an SOS-like signal to allow the activity of specific replication and repair functions necessary to minimize DNA damage. Carcinogen-induced amplification, overexpression, and viral induction can thus provide useful probes into the nature of such a putative pathway in higher eukaryotes.

REFERENCES

Aladjem, M., Koltin, Y., and Lavi, S. 1988. Enhancement of copper resistance and CupI amplification in carcinogen treated yeast cells. Mol. Gen. Genet. 211:88–94.

Alitalo, D., and Schwab, M. 1985. Oncogene amplification in tumor cells. Adv. Can. Res. 47:235–281.

Balmain, A., and Brown, K. 1988. Oncogene activation in chemical carcinogenesis. Adv. Can. Res. 51:147–183.

Benjamin, T., and Vogt, P. K. 1990. Cell transformation by viruses. In Virology, ed. B. N. Fields, pp. 317–381. New York: Raven Press.

Berko-Flint, Y. 1988. The molecular mechanism of carcinogen mediated SV40 DNA amplification—in vivo and in vitro studies. Ph.D. thesis, Weizmann Institute of Science.

Berko-Flint, Y., Karby, S., and Lavi, S. 1988. Carcinogen-induced factors responsible for SV40 DNA replication and amplification in Chinese hamster cells. Cancer Cells 6:183–189.

Berko-Flint, Y., Karby, S., Hassin, D., and Lavi, S. 1990. Carcinogen-induced DNA amplification in vitro: Overreplication of the simian virus 40 origin region in extracts from carcinogen-treated CO60 cells. Mol. Cell. Biol. 10:75–83.

Brown, K., Buchmann, A., and Balmain, A. 1990. Carcinogen induced mutations in the mouse c-Ha-ras gene provide evidence for multiple pathways for tumor progression. Proc. Natl. Acad. Sci. USA 87:538–542.

Cooper, C. S. 1990. The role of oncogene activation in chemical carcinogenesis. In Molecular Carcinogenesis and Mutagenesis II—Handbook of Experimental Pharmacology 94/2, eds. C. S. Cooper and P. L. Grover, pp. 319–351. Heidelberg: Springer-Verlag.

Defais, M. 1990. Mechanisms of repair in mammalian cells. In Molecular Carcinogenesis and Mutagenesis II—Handbook of Experimental Pharmacology 94/2, eds. C. S. Cooper and P. L. Grover, pp. 51–69. Heidelberg: Springer-Verlag.

Herrlich, P., Angel, P., Rahmsdorf, H. J., Mallick, U., Potig, A., Hieber, L., Lucke-Huhle, C., and Schorpp, M. 1986. The mammalian genetic stress response. Adv. Enzyme Regul. 25:485–505.

Johnston, R. N., Feder, J., Hill, A. B., Sherwood, S. W., and Schimke, R. T. 1986. Transient inhibition of DNA synthesis and subsequent increased DNA content per cell. Mol. Cell. Biol. 6:3373–3380.

Kleinberger, T., Etkin, S., and Lavi, S. 1986. Carcinogen mediated methotrexate resistance and dihydrofolate reductase amplification in Chinese hamster cells. *Mol. Cell. Biol.* 6:1958–1964.

Kleinberger, T., Berko-Flint, Y., Blank, M., Etkin, S., and Lavi, S. 1988a. Carcinogen induced trans activation of gene expression. *Mol. Cell. Biol.* 8:1366–1370.

Kleinberger, T., Sahar, E., and Lavi, S. 1988b. Carcinogen mediated co-activation of two independent genes in Chinese hamster cells. *Carcinogenesis* 9:979–985.

Lavi, S. 1981. Carcinogen-mediated amplification of viral DNA sequences in simian virus 40 transformed Chinese hamster embryo cells. *Proc. Natl. Acad. Sci. USA* 78:6144–6148.

Lavi, S., and Etkin, S. 1981. Carcinogen mediated induction of SV40 DNA synthesis in SV40 transformed Chinese hamster embryo cells. *Carcinogenesis* 2:417–423.

Li, J. J., and Kelly, T. J. 1985. Simian virus 40 replication in vitro: Specificity of initiation and evidence for bidirectional replication. *Mol. Cell. Biol.* 5:1238–1246.

Matz, B., Schlehofer, J., and zur Hausen, H. 1984. Identification of a gene function of HSV type I essential for amplification of SV40 sequences in transformed hamster cells. *Mol. Cell. Biol.* 4:1159–1162.

Melamed, D. 1988. Carcinogen-mediated activation of gene expression in Chinese hamster cells. M.Sc. thesis, Tel Aviv University.

Orth, G., Jablonska, S., Jarzabek-Chorzelska, M., Obalek, S., Rseza, G., Favre, M., and Croissant, O. 1979. Characteristics of the lesions and the risk of malignant conversion as related to the type of human papillomavirus involved in epidermodysplasia verruciformis. *Cancer Res.* 39:1074–1082.

Ronai, Z., Lambert, M. E., Johnson, M. D., Okin, E., and Weinstein, I. B. 1987. Induction of asynchronous replication of polyoma DNA in rat cells by UV irradiation and the effects of various inhibitors. *Cancer Res.* 47:4565–4570.

Schlehofer, J., Ehrber, M., and zur Hausen, H. 1986. Vaccinia virus, herpes virus and carcinogens induce DNA amplification in a human cell line and support replication of a helper virus dependent parvovirus. *Virology* 152:110–117.

Tlsty, T. D., Brown, P. C., Johnston, J., and Schimke, R. T. 1984. UV irradiation facilitates methotrexate resistance and amplification of the dihydrofolate reductase gene in cultivated 3T6 mouse cells. *Mol. Cell. Biol.* 4:1050–1056.

Valerie, C., Delers, A., Bruck, C., Thiriart, C., Rosenberg, H., Debouck, C., and Rosenberg, M. 1988. Activation of human immunodeficiency virus type I by DNA damage in human cells. *Nature* 333:78–81.

Yakobson, B., Hrynko, T. A., Peak, M. J., and Winocour, E. 1989. Replication of adeno associated virus in cells irradiated with UV light at 254 nm. *J. Virol.* 63:1023–1030.

9 | gadd153, A GROWTH ARREST AND DNA DAMAGE INDUCIBLE GENE: EXPRESSION IN RESPONSE TO GENOTOXIC STRESS

Nikki J. Holbrook, Jennifer D. Luethy, Jong Sung Park, Joseph Fargnoli

Laboratory of Molecular Genetics, Gerontology Research Center, National Institute on Aging, Baltimore, Maryland

INTRODUCTION

In *Escherichia coli,* treatments that damage DNA or inhibit its replication elicit the rapid and coordinate expression of some 20 different genes of metabolically diverse pathways, known collectively as the SOS response (Walker, 1985). As a result, a number of phenotypic changes occur, which include enhanced capacity for DNA repair and mutagenesis, inhibition of cell division, and prophage induction. While we do not know the functions of all of the genes activated during the SOS response, it is clear that it is an essential system for survival during adversity and is intimately linked to growth control.

Eukaryotes also respond to DNA damage with the induction of numerous genes. In yeast it has been predicted that more than 80 genes are activated by DNA damage (Ruby and Szostak, 1985); among these are genes linked to replication and growth (Elledge and Davis, 1989, 1990). Thus, based on the findings in both bacteria and yeast, it seems likely that a large number of genes would be induced by DNA damage in mammalian cells and that at least some of these would be linked to the control of cellular proliferation. A number of DNA damage inducible genes have been identified (for recent reviews see Kaina et al., 1990; Holbrook and Fornace, 1991). Table 1 lists the mammalian genes of which we are aware whose RNAs are increased in response to DNA damage. In addition to indicating their sensitivities to different types of DNA damage, we have also noted their responsiveness to phorbol ester stimulation and their growth-related expression. It should be pointed out that we have tried to be as complete as possible in denoting the specificity of the response. If an agent has been tested with no response it is so indicated. However, lack of an agent listed does not imply the response is negative, only that we do not know.

Few of these genes have been examined in depth. With the exception of β-polymerase (Miller and Chinault, 1982; Dresler and Lieberman, 1983), none of the genes (for which a function is known) appear to be involved in the actual repair of DNA, and for most the link to damage is

TABLE 1. DNA Damage Inducible Genes in Mammalian Cells

Gene	Inducing agent			
	UV	TPA	Other inducers	Growth
Collagenase[1,2]	+	+	MNNG, MMC, hydroxyurea, gamma radiation	>
Stromelysin[1]	+	+		>
Plasminogen activator[1,3]	+	+		>
Ornithine decarboxylase[4,5]	+	+		
c-fos[6,7]	+	+	MMS, MNNG, H_2O_2, heat shock	>
c-jun[1,8]	+	+	X-ray	>
c-myc[9]	+	+	X-ray	>
Egr-1[10]	+	+	X-ray	>
Metallothionein (MTEA)[2,11,12]	+	+	Glucocorticoid, heavy metals, MMG	
Heme oxygenase[13,14]	+	+	Arsenite, heavy metals, MMS, H_2O_2, others	
HIV 1[2,15]	+	+	NQO, MMC, not induced by MMS, X-rays	
SV40 promoter/enhancer[1,16]	+	+	MNNG, MMC	
MSV LTR[17]	+	+	X-rays	
sprI[18,19]	+	+	NQO	
sprII[18]	+	+	NQO	
spr2-1[19]	+	+	NQO	dif
DDI A class I[20]	+	−**		
DDI A class II[20]	+	−**		
gadd153[21,22]	+	−	MMS, many others but not X-rays	<
gadd45[21,23]	+	−	X-rays, MMS, others	<
gadd33, 34, 7[21]	+	−	MMS	<
β-Polymerase[24,25]	+	−	MMS, MNNG, AAAF, but not X-rays or bleomycin	
TNF[26,27]		+	X-rays	
IL-1[28,29]	+	+		>

Note. +, inducible; −, not inducible; **as tested in CHO cells after 4 h of treatment; >, induced by stimulation of growth; <, associated with growth arrest; dif, differentiation associated; UV, ultraviolet radiation, NQO, 4-nitroquinoline 1-oxide; MNNG, methyl-N'-nitro-N-nitrosoguanidine; MMC, mitomycin C; MMS, methyl methane sulfonate; AAAF, N-acetoxy-2-acetylaminofluorene; TPA, 12-O-tetradecanoyl-phorbol-13-acetate; MSV, Moloney sarcoma virus; HIV-1, human immunodeficiency virus I; *spr,* small proline rich; DDI, DNA damage inducible; TNF, tumor necrosis factor; IL-1, interleukin 1. References: (1) Kaina et al., 1990; (2) Stein et al., 1989; (3) Miskin and Ben Ishai, 1981; (4) Verma et al., 1979; (5) Rose-John et al., 1987; (6) Buscher et al., 1988; (7) Hollander and Fornace, 1989; (8) Devary et al., 1991; (9) Sullivan et al., 1989; (10) Hallahan et al., 1991; (11) Angel et al., 1986; (12) Hamer, 1986; (13) Applegate et al., 1991; (14) Keyse and Tyrrell, 1989; (15) Valerie et al., 1988; Valerie and Rosenberg, 1990; (16) Imbra and Karin, 1986; (17) Lin et al., 1990; (18) Kartasova and van de Putte, 1988; (19) Gibbs et al., 1990; (20) Fornace et al., 1988; (21) Fornace et al., 1989a; (22) Luethy et al., 1990; (23) Papathanasiou et al., 1991; (24) Fornace et al., 1989b; (25) Fornace, unpublished results; (26) Economou et al., 1989; (27) Hallahan et al., 1989; (28) Ansel et al., 1983; (29) Gahring et al., 1984.

not clear. However, as predicted, several of the genes are also growth regulated. From the available studies, at least two important facts have emerged. First, most of the DNA damage responsive genes are also induced by the tumor promoter 12-O-tetradecanoyl-phorbol-13-acetate (TPA). The *gadd* genes are a notable exception, as discussed in greater detail later. The close correlation between ultraviolet radiation (UV) and TPA responsiveness has led to speculation that protein kinase C is somehow involved in the DNA damage induced activation process, and evidence to support this notion has been found for several genes (Buscher et al., 1988; Kaina et al., 1990; Lin et al., 1990). Second, the specificity of the induction with respect to genotoxic agents varies for the different genes. Some genes respond only to UV or UV mimetic agents (most notably HIV-I), while others show a broader response to different kinds of DNA-damaging agents (*gadd153*, *gadd45*, and MSV LTR). Still others appear to represent a more generalized response to cell injury, as a variety of treatments, not directly damaging to DNA, will induce their expression (*c-fos*, *c-jun*, metallothionein). If we are to understand the complex mechanisms involved in controlling the eukaryotic stress response, the induced genes must be carefully examined to define the genetic basis for their responsiveness to DNA damage and to determine their relative sensitivity to different classes of genotoxic agents. Our own studies have concentrated primarily on the *gadd* genes as they appear to represent a unique class of DNA damage inducible transcripts.

THE *gadd* GENES: A UNIQUE CLASS OF GROWTH ARREST AND DNA-DAMAGE-INDUCIBLE GENES

The *gadd* genes were originally isolated from a hybridization subtraction library made from Chinese hamster ovary cells by Fornace et al. (1988). They were part of a larger group of some 20 cDNAs cloned on the basis of their rapid but modest induction following UV irradiation. The 20 genes were subsequently divided into two classes: class I transcripts were induced by UV radiation only, while class II transcripts were also induced by the DNA alkylating agent methyl methanesulfonate (MMS), and thus appeared to represent a more general response to DNA damage. Of the class II clones, five were found to be highly induced in a coordinate fashion by a variety of growth arrest signals, including medium depletion, contact inhibition, and serum starvation; these were named *gadd* (growth arrest DNA damage inducible) (Fornace et al., 1989a). The unique pattern of expression of the *gadd* genes is of particular interest as a well-known effect of DNA damage in both bacteria and eukaryotes is a transient inhibition of DNA synthesis and a delay in cell cycle progression. In bacteria, the *sulA* SOS response gene codes for a protein associated with growth arrest following DNA damage (Huisman et al., 1984), and the yeast RAD9 gene appears to be responsible, at least in part, for the delay of cell cycle progression following DNA damage (Weinert and Hartwell, 1988). It is possible that the *gadd* genes serve a similar function in mammalian cells.

None of the *gadd* cDNAs isolated from the subtraction library cross-hybridize. Southern analysis indicates that each is present as a single copy gene in the hamster genome. The full-length sequences for only two of the cDNAs, those for *gadd153* and *gadd45*, have been obtained. The *gadd153* cDNA sequence shares little homology to known sequences in the GenBank/EMBL Data Bank. However, analysis of the encoded protein indicates that *gadd153* is related to the CCAAT enhancer binding protein (C/EBP) gene family (discussed below). *gadd45* shares a high level of identity (>60%) with the mouse myeloid differentiation primary response gene MyD118, which was recently cloned from M1 myeloblastic leukemia cells based on its increased expression following induction of differentiation and growth arrest by IL6 (Abdollahi et al., 1991). Information for *gadd33*, *gadd34*, and *gadd7* has been based on the partial cDNA sequences obtained from the subtraction library, so it is possible that more sequence information will reveal homologies to known genes. Three of the *gadd* genes, *gadd153*, *gadd45*, and *gadd34*, were shown to be overexpressed in homozygous deletion c^{14cos}/c^{14cos} mice that are missing a small portion of chromosome 7 involving the albino locus (Glueckshon-Waelsch, 1979, 1987). This suggests the possibility that a common negative regulator of *gadd* gene expression is normally present in the deleted portion, but this has not been proven. The other two *gadd* genes, *gadd33* and *gadd7*, were not expressed in the mouse and may be specific to the hamster. *gadd153* and *gadd45* are ubiquitously expressed in human cells and are highly conserved among mammals (discussed later). The *gadd153* gene shows the highest level of induction in response to both DNA damage and growth arrest and has been studied most extensively. Here we summarize our studies examining the regulation of *gadd153* expression in response to DNA damage and growth arrest. We include information concerning *gadd45* where available, and attempt to discuss our findings with the *gadd* genes in relation to other DNA damage inducible genes.

STRUCTURE AND GENERAL FEATURES OF THE *gadd153* GENE AND ITS PROTEIN

We have reported the sequence and structure of the hamster *gadd153* gene (Luethy et al., 1990). It spans about 6 kb and contains 4 exons, separated by introns of 2.4, 1.8, and 0.17 kb in size. We have also recently cloned the human gene, which is even smaller in size, comprising only 3.5 kb due to the size of the introns (1.4, 0.3, and 0.09 kb) (Park et al., 1992). The overall gene structure is highly conserved from hamster to human, with exon/intron splice junctions conserved in identical positions in the two species. The hamster and human genes are transcribed into mRNAs of 837 and 867 base pairs (bp), respectively, and show high identity, particularly within the coding regions. A cDNA corresponding to the mouse homologue, referred to as CHOP-10 has also recently been isolated (Ron and Habener, 1992).

As mentioned above, the *gadd153* protein is related to the C/EBP family of transcriptional regulators which are generally expressed in a cell type-restricted manner, most notably in liver and adipocytes (Cao et al., 1991) and are also characterized by the presence of a conserved bZIP domain in the carboxyl terminal region of the protein (Vinson et al., 1989). This region consists of a basic domain involved in DNA sequence recognition and an adjacent helical "leucine zipper" domain through which dimerization of monomers occurs. *gadd153* is unique among the C/EBP related proteins in that it cannot bind to DNA (Ron and Habener, 1992). It can, however, form stable heterodimers with other C/EBP proteins and prevent their binding to DNA suggesting that *gadd153* acts as an inhibitor of these transcriptional activators. The functional significance of these interactions in vivo, and their relationship to the ubiquitous expression of *gadd153* in response to DNA damage and growth arrest is at present unclear.

An interesting feature of the human *gadd153* transcript is the presence of two adjacent AUUUA pentamers in the 3' untranslated portion of the mRNA. These sequences have been shown to be associated with instability of a number of mRNAs, including oncogenes, cytokines, and transcriptional factors, which are transiently expressed and have short half-lives (Brawerman, 1989). Accordingly, *gadd153* has a half-life of less than an hour in proliferating HeLa cells (unpublished results). Also within this region is the sequence TTATTTAT, which is particularly prevalent among mRNAs encoding proteins related to the inflammatory response (Caput et al., 1986).

THE *gadd153* PROMOTER AND ITS ACTIVATION
BY DNA DAMAGE

Using nuclear run-on experiments Fornace et al. (1989) demonstrated that the induction of *gadd153* mRNA (and also *gadd45*, *gadd33*, and *gadd34* mRNAs) by MMS and at least one growth arrest condition, that of medium depletion, occurred through an increase in the transcription of the gene. Thus we have been particularly interested in identifying the elements within the promoter region of *gadd153* that are responsible for its activation following DNA damage and/or growth arrest. Previously we published the sequence of 785 bp of the 5' region of the hamster *gadd153* gene (Luethy et al., 1990), and have since sequenced the promoter region of the human *gadd153* gene. Figure 1 shows a schematic comparison of the promoter regions of the hamster and human genes.

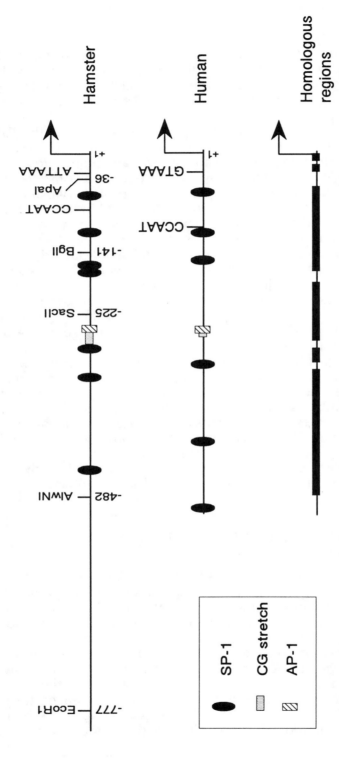

FIGURE 1. Hamster and human *gadd153* promoter regions share high identity. Schematic representation of the hamster (−800 to +21) and human (−500 to +21) *gadd153* 5′ flanking regions; +1 indicates the transcriptional start site. Regions of greater than 70% homology are shown with the bold line.

130

There is a high degree of identity within the 450-bp region proximal to the RNA start site. Regions sharing greater than 70% identity are shown; within these larger segments there are localized areas with 100% identity over 20- to 40-bp stretches. The promoters of both genes are relatively GC rich and contain a large number of putative SP-1 binding sites (GG-GCGG) (Kadonaga et al., 1987), two of which are positioned identically in the two genes. Of particular interest is the presence of an AP-1 binding site (Imbra and Karin, 1986) in an identical position in both genes with a series of alternating CpG residues immediately upstream of it. A number of other DNA damage inducible genes (collagenase, metallothionein, *spr2*-1, *spr2*-2) contain a similar AP-1 binding site. The high degree of conservation among the 450 bp of 5' flanking sequence in the human and hamster *gadd153* genes supports a role for these regions in mediating the response to DNA damage and/or growth arrest.

We have shown that the CHO *gadd153* promoter fragment from −778 to +21, when linked to the bacterial reporter gene chloramphenicol acetyltransferase (CAT), exhibits constitutive promoter activity, which can be further activated by treatment with UV irradiation, hydrogen peroxide, or MMS. To further explore the specificity of the *gadd153* promoter we have established a convenient reporter system in which the *gadd153* promoter linked to CAT is stably transfected into HeLa cells. Pools of stable transformants carrying the integrated chimeric gene have been examined for increases in CAT expression following their treatment with a variety of genotoxic as well as non-DNA-damaging but otherwise toxic compounds. A detailed study using this system has been published elsewhere (Luethy and Holbrook, 1992). As summarized in Table 2, CAT expression from the *gadd153* promoter is activated by a broad spectrum of DNA damaging agents. Of those tested, only x-rays and bleomycin failed to activate the *gadd153* promoter. This is consistent with our earlier observation that x-ray treatment did not alter *gadd153* mRNA levels of CHO cells (Fornace et al., 1989a). More recently, Papathanasiou et al. (1991) have confirmed this finding in human cells. Surprisingly, however, they found that *gadd45*, although not response to x-rays in CHO cells, was in fact induced by x-rays in several different human cell lines.

Attempts to categorize DNA damaging agents frequently have lumped them into two main groups: those that produce bulky DNA lesions and are repaired by the nucleotide excision repair systems [UV, mitomycin C, 4-nitroquinoline-*N*-oxide (NQO), and *N*-acetoxy-2-acetylaminofluorene], and those that result in nonbulky base damage and are repaired by base excision [x-ray, gamma radiation, alkylating agents (MMS), hydrogen peroxide, and topoisomerase inhibitors] (Regan and Setlow, 1974; Amacher et al., 1977; Sancar and Sancar, 1988). Our results and those of Papathanasiou et al. would indicate that this classification scheme is too general and that other factors need be considered. Characterization of different CHO and Chinese hamster V79 repair mu-

TABLE 2. Activation of the *gadd153* promoter by Toxic Substances in HeLa Cells

Agent	Dose	Time of treatment	Relative increase in expression[a]
Actinomycin D	0.25 µg/ml	4 h	+ + +
UV	30 J/m^2		+ + +
4-NQO[b]	100 ng/ml	24 h	+ + +
Mitomycin C	5 µg/ml	24 h	+ +
MMS	100 µg/ml	4 h	+ + +
Cisplatin	20 µg/ml	4 h	+ +
Hydrogen peroxide	600 µM	1 h	+ +
X-ray	800 R		−
Bleomycin	75 µg/ml	4 h	−
Arsenite	80 µM	2 h	+ +
Heat shock	42°C	2 h	−
Hydroxyurea	2 mM	24 h	+
TPA	30 ng/ml	24 h	−

Note. Exponentially growing *gadd153*CAT/HeLa cells were treated with the indicated agents for the length of time listed, after which the cells were washed and fed fresh medium. Cells were harvested 24 h later and CAT activities were determined.

[a]The relative increase in expression is determined as the ratio of CAT expression seen with and without treatment: + + +, >50-fold increase; + +, 10- to 50-fold increase; +, 3-fold increase; −, <2-fold increase in CAT expression.

[b]4-nitroquinoline-*N*-oxide.

tants and their cross-sensitivities to different kinds of DNA damaging agents likewise supports this view (Hickson and Harris, 1988).

DNA DAMAGE IS THE INDUCING SIGNAL FOR *gadd153* PROMOTER ACTIVATION

Activation of the *gadd153* promoter is not a general response to cell injury, since heat, which does induce the heat shock response, failed to influence CAT expression. Furthermore, an inhibition of DNA synthesis is not sufficient to induce the response, as hydroxyurea, which inhibits DNA synthesis and subsequently leads to growth arrest, did not significantly affect the expression of the *gadd153* CAT chimeric gene within the time frame of these experiments. Stein et al. (1989) and Valerie and Rosenberg (1990) have obtained similar results with the human collagenase and HIV-I promoters, respectively.

Such findings support the contention that DNA damage is the important signal for induction of *gadd153*. However, the strongest argument for a DNA target mediating the response is the dose-modifying effect seen in DNA repair-deficient cells. In cells stably transfected with *gadd-153*CAT, substantially smaller doses of UV radiation were required to produce an induction in excision-deficient (*uvr−*) xeroderma pigmentosum (XPA) cells equivalent to that observed in normal cells (Luethy and

Holbrook, 1992). While damage in non-DNA targets would be expected to be the same in normal and *uvr−* cells, DNA damage is not repaired and is more persistent in mutant cells in which *gadd153* is more strongly induced at lower UV radiation doses. *Gadd153*CAT expression in XPA cells does not differ from normal cells in response to MMS. This is not surprising, as XPA cells are not deficient in the repair of alkylation-induced DNA damage. Taken together, these results indicate that the signal for *gadd153* induction is damage to DNA.

To identify which regions of the *gadd153* promoter are responsible for its activation by DNA damage, we constructed a number of 5′ deletion mutants within the hamster promoter region between −778 and −36 and tested them for their responsiveness to MMS (Holbrook et al., 1991). These CAT constructs were derived from convenient restriction sites within the promoter region as indicated in Fig. 1. Maximum induction was seen with the full construct containing sequences to −778 (26-fold). Deletion to −482 resulted in about a 40% reduction in activation by MMS (16 fold); deletion to −225, which eliminates the AP-1 site, caused a further threefold decline in the response (fivefold). Further deletion to −141 resulted in only a slight decline in MMS responsiveness relative to the −225 deletion; MMS-induced CAT expression was still fourfold higher than that seen with a deletion to −36, which removes everything 5′ of the TATA site. We have tested several other DNA damaging agents, including NQO, cisplatin, and actinomycin D, for their effects on expression of some of these deletion constructs and have found they all behave similarly. From these studies we conclude that multiple regions within the 5′ flanking sequence contribute to the activation of the *gadd153* promoter by DNA damage and that the response to all DNA damage, regardless of the mode of action or pathway of repair, occurs through the same sequences.

THE AP-1 SITE IN THE *gadd153* PROMOTER IS DIFFERENTIALLY SENSITIVE TO MMS AND TPA

A general feature shared by DNA damage inducible genes is that they are also inducible by phorbol esters (Table 1). Indeed, in most cases they are more highly induced by the phorbol ester than by DNA damage. This regulation is transcriptional and is associated with the presence of a phorbol ester responsive element in the 5′ flanking regions of these genes. In the few cases thus far examined (HIV-I, *c-fos*, *c-jun*, and human collagenase), these phorbol ester responsive elements have likewise been shown to play a key role in the activation of gene expression following DNA damage (Stein et al., 1989; Devary et al., 1991). In particular, the human collagenase and *c-jun* genes both contain an AP-1 site that is critical for their regulation by DNA damage, as deletion or mutation of the site completely abolishes induction by UV irradiation. We originally reported that in CHO cells the *gadd* genes as a group were not sensitive to TPA, but this initial finding was restricted to a single cell type and to only two time points following TPA treatment (4 and 24 h). Therefore, we performed a more extensive survey to examine the influence of TPA on endogenous *gadd153* mRNA in other cell types (untransformed human diploid fibroblasts, Jurkat T-lymphocytes, and HeLa cells) and with extended time courses (0.5, 1.0, 2.0, 3.0, 6.0, 12.0, and 24.0 h). In no case have we observed more than a twofold increase in *gadd153* mRNA with TPA treatment. Consistent with these observations, the *gadd153* promoter was found to be relatively insensitive to TPA in all of these cell types (Luethy et al., 1990; also unpublished results).

As mentioned above, deletion of a 250-bp fragment (from -482 to -225; see Fig. 1) that encompasses the AP-1 binding site resulted in a threefold decrease in responsiveness of the *gadd153* promoter to DNA damage in HeLa cells. Because of this finding, and the relative importance of a similar AP-1 binding site in the DNA damage induced activation of the collagenase promoter, we performed additional experiments to determine whether the AP-1 site in the flanking region of *gadd153* gene is involved in the activation of *gadd153* expression by MMS. An oligonucleotide containing three copies of the genomic sequence CGCATGACTCACTCA was synthesized and linked to the basal *gadd153* promoter construct $-36/+21$. The construct, designated $(3 \times AP\text{-}1)-36/+21CAT$, was introduced into HeLa cells and tested for its expression relative to that of the full promoter construct, $-778/+21CAT$, following treatment of the transfected cells with MMS or TPA. Similar to the full-length native promoter $(-778/+21)$, expression of the $(3 \times AP\text{-}1)-36/+21CAT$ construct was significantly enhanced by MMS but not by TPA (Holbrook et al., 1991). Using a mobility shift assay we have demonstrated that HeLa extracts contain factors that specifically bind to this oligonucleotide sequence and that binding activity is increased in extracts from MMS-treated cells (Fig. 2). Based on competition assays this binding activity appears to be indistinguishable from that to other AP-1 binding sites (Fargnoli and Holbrook, unpublished results). Taken together these results suggest that the AP-1 site within the 5' flanking region of the *gadd153* gene does indeed contribute to the induction of *gadd153* expression following DNA damage.

Several studies have provided evidence that a protein kinase is involved in the DNA damage-induced expression of several genes. Herrlich and co-workers have shown that inhibitors of protein kinases block the UV-induced expression of genes at the mRNA level (Buscher et al., 1988). In addition, Lin et al. (1990) demonstrated that x-ray-, UV-, and TPA-induced CAT transcription from the Moloney sarcoma virus LTR in mouse 3T3 cells was blocked by treatment of the cells with the kinase inhibitor H7 (Papathanasiou et al., 1991). Evidence that the kinase was PKC was provided by the finding that pretreatment of cells with TPA to

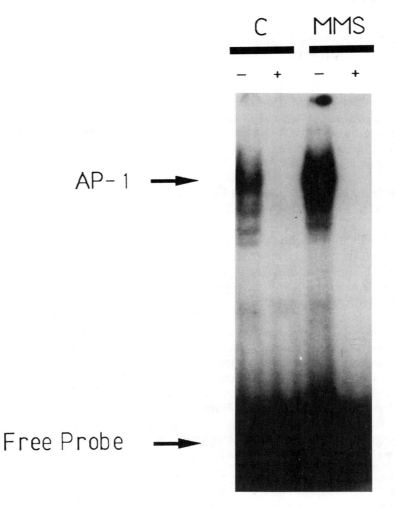

FIGURE 2. MMS treatment results in increased AP-1 binding activity in HeLa cells. Lysates from untreated (C) and MMS (100 µg/ml) treated HeLa cells were incubated with 1 ng of a ^{32}P-labeled oligonucleotide CGCA<u>TGACTCA</u>CTCA in the absence (−) or presence (+) of 10 ng unlabeled oligonucleotide.

down-regulate PKC also blocked the MSV response. Induction of *gadd153* and *gadd45* mRNAs by MMS is blocked by H7 (Papathanasiou et al., 1991) as well as by staurosporine and 2-aminopurine (unpublished findings) in a variety of cell types including HeLa. To determine if the protein kinase inhibitors exert their effects at the transcriptional level, we tested the ability of H7 and 2-aminopurine to prevent activation of the *gadd153* promoter in *gadd153*CAT/HeLa cells (Holbrook et al., 1991). Neither inhibitor alone had any effect on the promoter activity, nor did either appreciably affect the induction by MMS. Likewise, pretreatment of the cells with TPA, which alone resulted in a twofold increase in *gadd153* promoter-driven CAT expression over control levels, did not prevent activation by MMS but rather led to a twofold higher level of expression, indicating that the effects of TPA and MMS are distinct and additive. Thus, a protein kinase does not appear to be involved in the transcriptional activation of *gadd153*, but could play a role in the posttranscriptional regulation of the gene, perhaps by influencing the stability of *gadd153* transcripts.

REGULATION OF *gadd153* EXPRESSION DURING GROWTH ARREST

As indicated in Table 1, a number of DNA damage inducible genes are linked in a positive sense to cellular proliferation. Three of these, *c-fos*, *c-jun*, and *c-myc*, play a key role in mediating the proliferative response. In contrast, the *gadd* genes are negatively regulated by growth. Table 3 lists treatments (all of which result in either growth arrest or altered growth rates in HeLa cells) that we have examined for their influences on *gadd153* expression in HeLa cells. The results obtained with the different treatments vary, suggesting multiple levels of control. Growth of cells to high density without medium replacement results in high-level *gadd153* expression. In addition, placement of such medium taken from high-density cultures onto low-density proliferating cells will likewise cause induction of the gene. Both nuclear run-on experiments and CAT expression studies with stably transfected *gadd153*CAT/HeLa cells indicate that this effect occurs at the level of transcription. We are currently attempting to identify which DNA sequences are responsible for the

TABLE 3. Effect of Different Growth Arrest Conditions on *gadd153* Expression in HeLa Cells

Treatment	Increase in mRNA levels	Nuclear run on	Promoter activation
HDMD	>20-fold	+	>10-fold
CM	>20-fold	N.D.	>10-fold
PGA$_2$	>20-fold	−	No effect
− Glucose	>20-fold	N.D.	No effect
− Serum	No effect	N.D.	No effect

Note. Levels of expression were determined relative to those of untreated low-density proliferating cultures in complete medium with 10% fetal bovine serum. HDMD, high-density medium depleted; cells were grown to confluency and not refed for 2 d. CM, conditioned medium from HDMD cells; CM was placed on low-density proliferating cultures for 24 h. PGA$_2$, prostaglandin A$_2$; 10 μg/ml for 24 h. − Glucose, proliferating cells were fed fresh glucose-free medium with dialyzed serum. − Serum, cells fed medium with no serum up to 72 h. N.D., not determined.

transcriptional control. It will be particularly important to determine if the responsive elements overlap with those responsible for transcriptional activation following DNA damage. Surprisingly, treatment of HeLa cells with prostaglandin A_2 (PGA_2), which results in G_1 growth arrest (Ohno et al., 1988) and a marked increase in *gadd153* mRNA levels (Choi et al., 1992), has no effect on the transcriptional activity of the gene. The PGA_2-induced increase in *gadd153* mRNA levels is prevented by the same protein kinase inhibitors that block MMS-induced mRNA expression. Thus, it is likely that PGA_2 acts at the posttranscriptional level to enhance *gadd153* mRNA expression and that this effect is dependent on a protein kinase. Glucose deprivation also elevated *gadd153* mRNA levels, again without influencing promoter activity. While serum deprivation causes an increase in *gadd153* expression in CHO cells, it is without effect in HeLa cells. However, these cells are not growth arrested (as are CHO cells) in the absence of serum, but merely grow at a slower rate. This most likely accounts for the lack of *gadd153* expression under these conditions. Thus, the growth arrest-induced expression of *gadd153* is complex and involves both transcriptional and posttranscriptional controls, depending on the particular conditions used to achieve arrest.

SUMMARY

The *gadd* genes offer an excellent opportunity to shed light on the molecular mechanisms involved in the mammalian response to genotoxic stress. Compared to other DNA damage inducible genes that have been examined in mammalian cells, *gadd153* is unique in several respects. It is not inducible by phorbol esters, and its transcriptional activation does not appear to be PKC dependent, despite the presence of an AP-1 binding site in the promoter region that is involved in its regulation by DNA damage. The mechanisms involved in the transcriptional activation of *gadd153* by DNA damage and the role of the AP-1 binding sequence in determining the differential response to DNA damage and TPA are currently under investigation. Likewise, studies are underway to determine the significance of the *gadd153* protein interactions with the C/EBP family of transcriptional regulatory proteins. Unlike other higher eukaryotic DNA damage inducible genes, the *gadd* genes are regulated by growth arrest conditions, and therefore likely play a fundamental role in normal cellular homeostasis, as has been shown for some of the DNA damage inducible gene functions in bacteria and yeast. Whether they play a role in initiating and/or maintaining the state of growth arrest is not clear. Delineation of the regulatory pathways involved in *gadd* gene expression and elucidation of the function of their encoded proteins should provide answers to these questions and increase our understanding of the mammalian response to genotoxic stress.

REFERENCES

Abdollahi, A., Lord, K. A., Hoffman-Liebermann, D. A. 1991. Sequence and expression of a cDNA encoding MyD118: A novel myeloid differentiation primary response gene induced by multiple cytokines. *Oncogene* 6:165–167.

Amacher, D. E., Elliot, J. A., and Lieberman, M. W. 1977. Differences in removal of acetylaminofluorene and pyrimidine dimers from the DNA of cultured mammalian cells. *Proc. Natl. Acad. Sci. USA* 74:1553–1557.

Angel, P., Poting, A., Mallick, U., Rahmsdorf, H. J., Schorpp, M., and Herrlich, P. 1986. Induction of metallothionein and other mRNA species by carcinogens and tumor promoters in primary human skin and fibroblasts. *Mol. Cell. Biol.* 6:1760–1766.

Ansel, J. C., Luger, T. A., and Green, I. 1983. The effect of *in vitro* UV irradiation on the production of ETAF activity by human and murine Keratinucyte. *J. Invest. Dermatol.* 81:519–523.

Applegate, L. A., Luscher, P., and Tyrrell, R. M. 1991. Induction of hemeoxygenase: A general response to oxidant stress in cultured mammalian cells. *Cancer Res.* 51:974–978.

Brawerman, G. 1989. mRNA decay: Finding the right targets. *Cell* 57:9–10.

Buscher, M., Rahmsdorf, H. J., Liftin, M., Karin, M., and Herrlich, P. 1988. Activation of *c-fos* gene by UV and phorbol ester: Different signal transduction pathways converge to the same enhancer element. *Oncogene* 3:301–311.

Cao, Z., Umek, R. M., and McKnight, S. L. 1991. Regulated expression of three C/EBP isoforms during adipose conversion of 3T3-L1 cells. *Genes and Develop.* 5:1538–1552.

Caput, D., Beutler, B., Hartog, K., Thayer, R., Brown-Shimer, S., and Cerami, A. 1986. Identification of a common nucleotide sequence in the 3'-untranslated region of mRNA molecules specifying inflammatory mediators. *Proc. Natl. Acad. Sci. USA* 83:1670–1674.

Choi, A. M. K., Fargnoli, J., Carlson, S. G., and Holbrook, N. J. 1992. Cell growth inhibition by prostaglandin A$_2$ results in elevated expression of *gadd153* mRNA. *Exp. Cell Res.* 199:85–89.

Chou, P. Y., and Fasman, G. D. 1978. Empirical predictions of protein conformations. *Annu. Rev. Biochem.* 47:251–276.

Devary, Y., Gottlieb, R. A., Lau, L. F., and Karin, M. 1991. Rapid and preferential activation of the c-jun gene during the mammalian UV response. *Mol. Cell. Biol.* 11:2084–2811.

Dresler, S. L., and Lieberman, M. W. 1983. Identification of DNA polymerases involved in DNA excision repair in diploid human fibroblasts. *J. Biol. Chem.* 258:9990–9994.

Economou, J. S., Rhoades, K., Essner, R., McBride, W. H., Gasson, J. C., and Morton, D. L. 1989. Genetic analysis of the human tumor necrosis factor, factor alpha/catechin promoter region in a macrophage cell line. *J. Exp. Med.* 170:321–326.

Elledge, S. J., and Davis, R. W. 1989. DNA damage induction of ribonucleotide reductase. *Mol. Cell Biol.* 9:4932–4940.

Elledge, S. J., and Davis, R. W. 1990. Two genes differentially regulated in the cell cycle and by DNA-damaging agents encode alternative regulatory subunits of ribonucleotide reductase. *Genes Dev.* 4:740–751.

Fornace, A. J., Jr., Alamo, I., Jr., and Hollander, M. C. 1988. DNA damage-inducible transcripts in mammalian cells. *Proc. Natl. Acad. Sci. USA* 85:8800–8804.

Fornace, A. J., Jr., Nebert, D. W., Hollander, M. C., Luethy, J. D., Papathanasiou, M., Fargnoli, J., and Holbrook, N. J. 1989a. Mammalian genes coordinately regulated by growth arrest signals and DMA-damaging agents. *Mol. Cell. Biol.* 9:4196–4203.

Fornace, A. J., Jr., Zmudka, B. Z., Hollander, M. C., and Wilson, S. H. 1989b. Induction of polymerase mRNA by DNA damaging agents in Chinese hamster ovary cells. *Mol. Cell. Biol.* 9:851–853.

Gahring, L., Baltz, M., Pepys, M. B., and Daynes, R. 1984. Effect of UV radiation on production of epidermal cell thymocyte-activating factor (interleukin) in vivo and in vitro. *Proc. Natl. Acad. Sci. USA* 81:1198–1202.

Garnier, J., Osgulthorpe, D. J., and Robsoro, B. 1978. Analysis of the accuracy and implications of simple methods for predicting the secondary structure of globular proteins. *J. Mol. Biol.* 120:97–120.

Gibbs, S., Lohman, F., Teabel, W., van de Putte, P., and Backendorf, C. 1990. Characterization of the human *spr2* promoter: Induction after UV irradiation or TPA treatment and regulation during differentiation of cultured keratinocytes. *Nucleic Acids Res.* 18:4401–4407.

Gluecksohn-Waelsch, S. 1979. Genetic control of morphogenetic and biochemical differentiation: Lethal albino deletions in the mouse. *Cell* 16:225–237.

Gluecksohn-Waelsch, S. 1987. Regulatory genes in development. *Trends Genet.* 3:123–127.

Hallahan, D. E., Sprigg, D. R., Beckett, M. A., Kufe, D. W., and Weichselbaum, R. R. 1989. Increased tumor necrosis factor mRNA after cellular exposure to ionizing radiation. *Proc. Natl. Acad. Sci. USA* 86:10104–10107.

Hallahan, D. E., Sukhatme, V. P., Sherman, M. L., Virudachalam, S., Kufe, D., and Weichselbaum, R. R. 1991. Protein kinase C mediates x-ray inducibility of nuclear signal transducers EGR1 and JUN. *Proc. Natl. Acad. Sci. USA* 88:2156–2160.

Hamer, D. H. 1986. Metallothionein. *Annu. Rev. Biochem.* 55:913–951.

Hickson, I. D., and Harris, A. L. 1988. Mammalian DNA repair-use of mutants hypersensitive to cytotoxic agents. *Trends Genet.* 4:101–106.

Holbrook, N. J., and Fornace, A. J., Jr. 1991. Response to adversity: Molecular control of gene activation following genotoxic stress. *New Biol.* 3:825–833.

Holbrook, N. J., Leuthy, J. D., Park, J. S., and Fargnoli, J. 1991. Molecular characterization of *gadd153,* a novel DNA damage-inducible gene in mammalian cells. In *Oxidative Damage and Repair: Chemical Biological and Medical Aspects,* ed. K. J. A. Davies. New York: Pergamon Press.

Hollander, M. C., and Fornace, A. J., Jr. 1989. Induction of *c-fos* RNA by DNA damaging agents. *Cancer Res.* 49:1687–1692.

Huisman, O., D'Ari, R., and Gottesman, S. 1984. Cell-division control in *Escherichia coli*: Specific induction of the SOS function sfiA protein is sufficient to block septation. *Proc. Natl. Acad. Sci. USA* 81:4490–4494.

Imbra, R. J., and Karin, M. 1986. Phorbol ester induces the transcriptional stimulatory activity of the SV40 enhancer. *Nature* 323:555–558.

Kadonaga, J. T., Carner, K. R., Masiarz, F. R., and Tijan, R. 1987. Isolation of cDNA encoding transcription factor sp1 and functional analysis of the DNA binding domain. *Cell* 51:1079–1090.

Kaina, B., Stein, B., Schonthal, A., Rahmsdorf, H. J., Ponta, H., and Herrlich, P. 1990. An update of the mammalian UV response: Gene regulation and induction of a protective function. In *DNA Repair Mechanisms and their Biological Implications in Mammalian Cells,* eds. M. W. Lamkert and J. Laval, pp. 149–165. New York: Plenum.

Kartasova, T., and van de Putte, P. 1988. Isolation, characterization, and UV-stimulated expression of two families of genes encoding polypeptides of related structure in human epidermal keratinocytes. *Mol. Cell. Biol.* 8:2195–2203.

Keyse, S., and Tyrrell, R. M. 1989. Heme oxygenase is the major 32-kDa stress protein inducted in human skin fibroblasts by UVA radiation, hydrogen peroxide, and sodium arsenite. *Proc. Natl. Acad. Sci. USA* 86:99–103.

Lin, C. S., Goldthwait, D. A., and Samols, D. 1990. Induction of transcription from the long terminal repeat of Moloney murine sarcoma provirus by UV-irradiation, X-ray irradiation and phorbol ester. *Proc. Natl. Acad. Sci. USA* 87:36–40.

Luethy, J. D. ,and Holbrook, N. J. 1992. Activation of the *gadd153* promoter by genotoxic agents: A rapid and specific response to DNA damage. *Cancer Res.* 52:5–10.

Luethy, J. D., Fargnoli, J., Park, J. S., Fornace, A. J., Jr., and Holbrook, N. J. 1990. Isolation and characterization of the hamster *gadd153* gene. *J. Biol. Chem.* 265:16521–16526.

Miller, M. R., and Chinault, D. N. 1982. The roles of DNA polymerases α, β, γ, and in DNA synthesis induced in hamster and human cells by different DNA damaging agents. *J. Biol. Chem.* 257:10204–10209.

Miskin, R., and Ben-Ishai, R. 1981. Induction of plasminogen activator by UV light in normal and xeroderma pigmentosum fibroblasts. *Proc. Natl. Acad. Sci. USA* 78:6236–6240.

Ohno, K., Sakai, T., Fukushima, M., Narumiya, S., and Fujiwara, M. 1988. Site and mechanism of growth inhibition by prostaglandins. IV. Effect of cytopentenone prostaglandins on cell cycle progression of G, enriched HeLa 53 cells. *J. Pharmacol. Exp. Ther.* 4:294–298.

Papathanasiou, M. A., Kerr, N. C., Robbins, J. H., McBride, O. W., Alamo, I., Jr., Barrett, S. F., Hickson, I. D., and Fornace, A. J., Jr. 1991. Induction by ionizing radiation of the *gadd45* gene in cultured human cells: Lack of mediation by protein kinase C. *Mol. Cell. Biol.* 11:1009–1016.

Park, J. S., Luethy, J. D., Wang, M. G., Fargnoli, J., Fornace, A. J., Jr., McBride, O. W., and Holbrook, N. J. 1992. Isolation characterization and chromosomal localization of the human *GADD153* gene. *GENE* (in press).

Regan, J. D., and Setlow, R. B. 1974. Two forms of repair in the DNA of human cells damaged by chemical carcinogens and mutagens. *Cancer Res.* 34:3318–3325.

Ron, D., and Habener, J. F. 1992. CHOP, a novel developmentally regulated nuclear protein that dimerizes with transcription factors C/EBP and LAP and functions as a dominant-negative inhibitor of gene transcription. *Genes and Develop.* 6:439–453.

Rose-John, S., Rincke, G., and Marks, F. 1987. The induction of ornithine decarboxylase by the tumor promoter TPA is controlled at the post-transcriptional level in murine Swiss 3T3 fibroblasts. *Biochem. Biophys. Res. Commun.* 147:219–225.

Ruby, S. W., and Szostak, J. W. 1985. Specific *Saccharomyces cerevisiae* genes are expressed in response to DNA-damaging agents. *Mol. Cell. Biol.* 5:75–84.

Sancar, A., and Sancar, G. B. 1988. DNA repair enzymes. *Annu. Rev. Biochem.* 57:29–67.

Stein, B., Rahmsdorf, H. J., Steffen, A., Liftin, M., and Herrlich, P. 1989. UV-induced DNA damage is an immediate step in UV-induced expression of human immunodeficiency virus Type I, collagenase, *c-fos*, and metallothionein. *Mol. Cell. Biol.* 9:5169–5181.

Sullivan, N. F., and Willis, A. E. 1989. Elevation of *c-myc* protein by DNA strand breakage. *Oncogene* 4:1497–1502.

Valerie, K., and Rosenberg, M. 1990. Chromatin structure implicated in activation of HIV-I gene expression by ultraviolet light. *New Biol.* 2:712–718.

Valerie, K., Delers, A., Bruck, C., Thiriart, C., Rosenberg, H., Debouck, C., and Rosenberg, M. 1988. Activation of human immunodeficiency virus type 1 by DNA damage in human cells. *Nature* 333:78–81.

Verma, A. K., Lowe, N. J., and Boutwell, R. K. 1979. Induction of mouse epidermal ornithine decarboxylase activity and DNA synthesis by ultraviolet light. *Cancer Res.* 39:1035–1040.

Vinson, C. R., Sigler, P. B., and McKnight, S. L. 1989. Scissors-grip model for DNA recognition by a family of leucine zipper proteins. *Science* 246:911–916.

Walker, G. C. 1985. Inducible DNA repair systems. *Annu. Rev. Biochem.* 54:425–457.

Weinert, T. A., and Hartwell, L. H. 1988. The RAD9 gene controls the cell cycle response to DNA damage in *Saccharomyces cerevisiae*. *Science* 241:317–322.

10 | ULTRAVIOLET IRRADIATION AND PHORBOL ESTERS INDUCE GENE TRANSCRIPTION BY DIFFERENT MECHANISMS

Hans J. Rahmsdorf, Stephan Gebel, Marcus Krämer, Harald König, Christine Lücke-Huhle, Adriana Radler-Pohl, Christoph Sachsenmaier, Bernd Stein, Hans-Peter Auer, Mirko Vanetti, and Peter Herrlich

Kernforschungszentrum Karlsruhe, Institut für Genetik und Toxikologie, Karlsruhe, Germany

INTRODUCTION

In certain mouse strains two classes of xenobiotics, initiators and promoters, lead synergistically to the formation of skin carcinomas. In this process initiators such as ultraviolet irradiation (UV), x-rays, or 7,12-dimethylbenz[a]anthracene are believed to cause mutations in certain regulatory genes, such as proto-oncogenes and presumably tumor suppressor genes, both of which normally control the cell cycle; promoters are thought to cause epigenetic changes that push the initiated cells into continuous proliferation, thus perhaps increasing the risk of further genetic changes such as chromosomal loss, deletions, and replication errors (for review see Diamond, 1987). In contrast to promoters, which, when given alone, lead only rarely to tumor formation, initiators at doses higher than those used in the initiation/promotion protocol regularly induce papillomas and carcinomas, suggesting that they also possess promoting activity. Moreover, experiments with cell cultures have suggested that initiators induce cellular transformation under conditions that are highly unlikely to cause direct mutations in regulatory genes (e.g., Kennedy et al., 1980). Rather, the initiator seems to establish in all cells a stable situation in which the probability of a subsequent mutation in a regulatory gene is higher than in control cells. Tumor promoters, it appears, do not induce this state, but enhance the efficiency of initiators (e.g., Kennedy et al., 1978). Thus, tumor promoters and initiators such as UV, although partly overlapping in their induced reactions, must address some targets that are clearly different.

This putative difference has given the motivation to study and compare the genetic reactions induced by UV and by phorbol esters in cells in culture. We chose UV-C irradiation and 12-O-tetradecanoyl-phorbol-13-acetate (TPA) as the inducing agents. As end points we examined gene transcription and gene replication. Over the first few years of this work, the similarity between the cellular reactions to the initiator and the promoter seemed overwhelming. At present we are detecting differences. In this chapter we focus on some of the differences in the reaction of cells to UV and to TPA. We wish to stress at this point that the differences do not yet solve, in any obvious way, why treatment of cells with UV, but not with TPA, leads to a fully transformed phenotype.

UV AND TPA INDUCE SIMILAR CHANGES IN THE PROGRAM OF GENE EXPRESSION AND GENE REPLICATION

When we asked some 10 years ago whether UV irradiation induced, as it does in *Escherichia coli,* the synthesis of new proteins in mammalian cells, we also examined the phorbol ester TPA for its influence on mammalian protein synthesis. We were surprised that the two agents induced similar changes in the pattern of proteins synthesized in primary human fibroblasts: the synthesis of at least seven proteins was switched on and the synthesis of at least one protein was repressed (Mallick et al., 1982). One of the induced proteins turned out later to be identical with collagenase I (Whitham et al., 1986). Since then, many laboratories, including our own, have greatly expanded the list of UV and TPA inducible proteins. A recent count yielded 83 phorbol ester-induced genes and around 30 genes induced by UV-C (Rahmsdorf and Herrlich, 1990; Herrlich et al., 1992). The expression of several genes is activated by both agents: for example, genes coding for transcription factors and replication factors (*c-jun, junB, c-fos, c-myc*), as well as genes coding for growth factors (interleukin 1α, tumor necrosis factor α) and for secreted proteases (collagenase I, stromelysin, plasminogen activator). Also, the genes coding for metallothioneins and several viral genomes are induced by both agents. This list may grow rapidly: many of the TPA-inducible genes have not yet been examined for UV inducibility and vice versa. The interest in the UV-induced genetic response in mammalian cells is catching up.

Besides induction of expression of specific genes, another common consequence of treatment of cells with UV or with tumor promoters is an "attempt" to proliferate. The UV pushes quiescent primary human fibroblasts into a round of DNA synthesis similar to the one induced by growth factors (Cohn et al., 1984). TPA also stimulates resting fibroblasts to divide (e.g., Driedger and Blumberg, 1977). Moreover, both UV and phorbol esters induce the overreplication of specific genes (gene amplification), as measured for instance by the frequency with which methotrexate-resistant cells (which have amplified their dihydrofolate reductase gene) appear during selection with this drug (Varshawski, 1981; Tlsty et al., 1984; see also chapters by Lücke-Huhle, Ronai et al., and Aladjem and Lavi, this volume). We come back to differences in UV- and TPA-induced gene amplification later.

COMMON FEATURES OF UV- AND TPA-INDUCED GENE EXPRESSION

The synthesis of a protein can be influenced on many levels, such as the rate of transcription, the stability of the messenger RNA, or the rate of translation of the messenger. In most cases examined, both phorbol esters and UV increased the transcription of the respective genes. This was shown by "nuclear run on" experiments and by examining the inducibility of hybrid genes in which the promoter of a responsive gene controlled the transcription of the reporter gene coding for chloramphenicol-acetyltransferase. These latter experiments also allowed examination of the features that distinguish a responsive from an unresponsive promoter. Similar experiments performed earlier with, for example, genes inducible by glucocorticoid hormone, by heavy metal or heat shock had shown that such genes carry recognition elements (e.g., enhancers) specific for the inducer and shared by genes that are activated by the same signal. The recognition elements are the binding sites for specific transcription factors. Several glucocorticoid-responsive genes, such as the mouse mammary tumor virus long-terminal repeat, the metallothionein IIA, or the tyrosine amino transferase gene, possess a DNA element that is recognized by the glucocorticoid receptor (Evans, 1988); different heavy-metal-responsive genes carry in their promoters the metal response element (Karin et al., 1987); and different heat-shock-responsive genes carry the heat-shock response element (Sorger, 1991). A gene such as the human metallothionein IIA gene that responds to two different inducers (glucocorticoids and heavy metals) carries in its promoter both enhancer elements (Karin et al., 1984).

Quite to our surprise, the situation turned out to be different in many of the phorbol ester and UV inducible genes. While as in the examples just cited, UV- and phorbol ester-inducible genes contain in their promoters specific enhancer sequences, in three of four genes examined (c-fos, human collagenase, and HIV-1 LTR), UV and TPA acted through the same sequence within the respective promoter. "Only in the c-jun promoter, the two UV response elements could be destroyed without affecting phorbol ester inducibility of the gene constructs" (Stein et al., 1992). Moreover, and in contrast to the situation with hormones, heavy metals, or heat shock, the responding sequence differed from one gene to the other: In the case of the c-fos promoter, the induction was mediated through the major enhancer (the dyad symmetry element, DSE) located between positions −300 and −320 and, to a minor extent, through a protein binding site located within the transcribed but untranslated region of the gene (Büscher et al., 1988). Even more confusingly, this latter site has been shown to react also to cyclic adenosine monophosphate (Härtig et al., 1991). The DSE binds several transcription factors, the serum response factor and two 62-kD proteins; the protein(s) binding to the transcribed region is (are) not yet known. In the case of the collagenase gene, the major site responding to both UV and phorbol esters is the AP-1 (c-Fos/c-Jun) binding site localized between positions −72 and −66 (Angel et al. ,1987a, 1987b; Stein et al., 1989a; Jonat et al., 1990); in the case of the HIV-1 LTR, the UV- and TPA-regulated element binds the transcription factor NFκB (Baeuerle and Baltimore, 1988a, 1988b; Stein et al., 1989a). Thus, in contrast to hormones, heavy metals, and heat shock, the carcinogens TPA and UV activate not one but several different transcription factors. In the case of AP-1 (c-Fos/c-Jun) and NFκB, both agents lead to an enhanced binding to their respective enhancers (Stein et al., 1989a). The phorbol ester- and UV-inducible enhancers are also controlled by physiological modulators of gene expression: growth factors. The dyad symmetry element of the c-fos promoter had been recognized before as the element of the c-fos gene that mediates the response to serum growth factors (Treisman, 1985). Also, the enhancers of the collagenase gene and the HIV-1 LTR respond to growth factors. Obviously, cells did not develop a complete new set of signal components to respond to phorbol esters and to UV. Rather these agents feed into physiological pathways.

The UV- or TPA-induced enhancement of activity of a transcription factor could be due either to enhanced synthesis of the protein or to activation of the protein by posttranslational modification. Because activation of transcription in many cases (c-fos, c-jun, HIV-1, gene constructs containing the AP-1 binding site and a reporter) also occurs in the absence of protein synthesis (Krämer et al., 1990), posttranslational modification of transcription factors must play the decisive role. Induction of the endogenous collagenase gene and of some other genes activated only late after the treatment of cells, is reduced in the presence of inhibitors of protein synthesis (Angel et al., 1987a; Krämer et al., 1990; Holbrook and Fornace, 1991; also see later in this chapter). Because the inducibility of the collagenase promoter is enhanced by sequences upstream of the AP-1 binding site (Angel et al., 1987a; Stein et al., 1989a; Jonat et al., 1992), the reduction of collagenase induction by inhibitors of protein synthesis could be due either to the reduced synthesis of proteins binding to these upstream sequences or to the inhibition of AP-1 (c-Fos/c-Jun) synthesis.

Phosphorylation and dephosphorylation reactions are efficient tools to modulate the activity of transcription factors. Many growth-factor receptors possess tyrosine kinase activity, which is activated upon binding of growth factors; the receptor for phorbol esters is the cytoplasmic protein kinase C, which upon binding to TPA is translocated to the cellular membrane and activated (Nishizuka, 1984). The protein kinase activities of these receptors are important for signal transduction to the responsive genes: protein kinase inhibitors interfere with the signal transduction pathways. The same has been found for the UV-induced transcription of genes. The protein kinase inhibitor H7 inhibits both phorbol ester- and UV-induced *c-fos* transcription (Büscher et al., 1988; for a discussion of the effect of H7 on other DNA damage inducible genes see Holbrook and Fornace, 1991). In an attempt to find inhibitors specific for either UV or TPA, we were unsuccessful: The inhibitors examined influenced UV and phorbol ester induced *c-fos* transcription similarly, suggesting that the same protein kinase (or protein kinases with similar properties) is involved in the two signal chains (Krämer et al., 1990).

Many of the genes and gene products stimulated by UV or TPA are themselves cellular oncogenes. Their uncontrolled activation transforms cells. Cells combat this by a number of safeguards. One of these is counterregulation. Induced gene transcription is limited, and some of the oncogene-derived mRNAs and proteins show extremely short half-lives (Kovary and Bravo, 1991). An interesting feature is that one and the same inducer cannot turn on the same gene twice. Cells are refractory for considerable periods of time (more than 24 h). For instance, in cells in which the *c-fos* gene has been induced by phorbol esters, a second treatment with this same drug is inefficient. UV-treated cells respond with *c-fos* transcription. A second treatment, however, does not cause an induction (Büscher et al., 1988). These findings suggest that both inducers use exhaustible components. We return to this phenomenon later.

DIFFERENCES IN THE UV- AND PHORBOL ESTER-INDUCED GENETIC PROGRAM

The aforementioned data suggested far-reaching identity in the genetic programs induced by UV and phorbol esters, and in the induction mechanisms. More recent experiments have changed our view. Differences have become evident, particularly in the kinetics of gene induction, in the responses of different cells and even in the mechanism of transcriptional activation of genes.

In Primary Human Fibroblasts UV-Induced Collagenase mRNA Accumulation is Delayed as Compared to the Phorbol Ester-Induced Process

In contrast to the rapid response of a gene construct containing only the AP-1 binding site of the collagenase gene within a minimal promoter linked to a reporter (Stein et al., 1989a), the endogenous collagenase gene is turned on with considerable delay. The length of this delay differs between phorbol ester- and UV-treated cells (Fig. 1): In phorbol ester-treated cells accumulation of collagenase mRNA is detectable as early as 1 h after treatment. The UV-induced collagenase mRNA can only be detected at 12–24 h after irradiation. As both inducers stimulate AP-1 synthesis with similar kinetics and to a similar extent (Fig. 1), a lack of

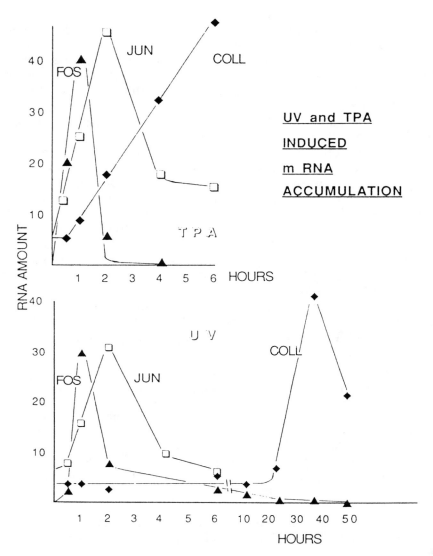

FIGURE 1. Different kinetics of collagenase mRNA induction by UV and by TPA. Primary human fibroblasts from a healthy donor were grown at 37°C and 6% CO_2 in Dulbecco's minimal essential medium supplemented with 15% fetal calf serum (FCS), 100 U/ml penicillin, 100 μg/ml streptomycin. Prior to the experiment the cells were starved in DMEM containing 0.5% FCS for 36 h. Cells were irradiated with UV (254 nm, 30 J/m^2) or treated with TPA (60 ng/ml); poly A$^+$ RNA was prepared and analyzed for c-fos, c-jun, and collagenase-specific mRNA as described by Rahmsdorf et al. (1987). Autoradiograms of the dried gels were scanned with a laser densitometer; arbitrary units are plotted.

cellular AP-1 cannot explain this difference in the capacity to induce the transcription of the collagenase gene.

In Several Cells the Immediate Early Genes c-fos and c-jun are Differently Regulated by UV and by Phorbol Esters

In HeLa cells both UV and TPA induce c-jun RNA accumulation with similar rapid kinetics and independently of ongoing protein synthesis. In contrast to the situation with primary human fibroblasts (Fig. 1), in HeLa cells UV is about three to four times more efficient as compared to phorbol esters (Devary et al., 1991; Stein et al., 1992). This is the reverse of the situation with c-fos, which is more efficiently induced by TPA (Devary et al., 1991). This difference in c-jun activation is reproducible in transient transfection experiments with c-jun promoter-CAT-constructs. For instance, a $-1600/+170$ c-jun CAT construct is induced around sixfold by UV and only twofold by TPA. The UV induction depends largely on the presence of two protein binding sites, one located between positions -71 and -64 and the other between positions -190 and -183. Both sites bind factors belonging to the AP-1 family, but are not identical to c-Fos/c-Jun (Stein et al., 1989b, 1992). If both sites are destroyed by insertional mutagenesis, UV induction of the promoter-CAT construct drops to twofold (Stein et al., 1992). Phorbol ester induction of this gene construct, however, is not changed, suggesting the existence of other responsive sites in the promoter. A 5' deletion mutant, $-196/+70$ CAT, behaves like the long promoter construct. Deletion to -160, however, reduced UV inducibility from sixfold to threefold, whereas phorbol ester inducibility was increased from twofold to sixfold (Stein et al., 1992). As isolated elements cloned in front of a TATA box, both the $-190/-183$ element and the $-71/-64$ element are induced to a similar extent by UV and by TPA (Stein et al., 1992). Obviously, it is dangerous to draw conclusions on the regulation of promoters by analyzing short sequences apart from their sequence context.

As in most other cells, in 3T3 cells UV and TPA induce the rapid and transient transcription of c-fos. At 2 h after treatment, the c-fos mRNA level is back to control levels (Fig. 2, upper panel). We found to our surprise that at later times after UV irradiation (but not after phorbol ester treatment), c-fos RNA accumulated again (Fig. 2). It is not known whether this second burst of c-fos RNA accumulation is due to enhanced transcription of the gene or to mRNA stabilization and whether there is a relationship to UV-induced growth factor secretion (see later).

It is generally accepted that a cell will respond to a growth factor if it carries the growth-factor receptor. Similarly, progesterone will act only on cells that possess the progesterone receptor. What about the UV response? Are there cells that lack a component of the DNA damage-induced signaling? Such cells have yet not been detected. Interestingly,

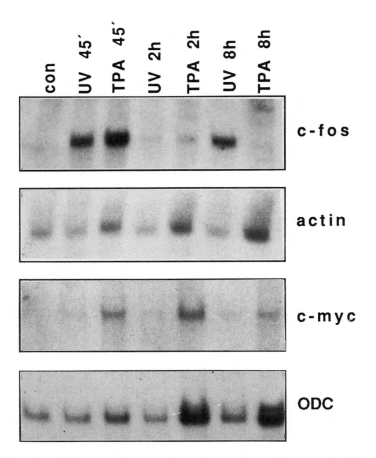

FIGURE 2. The c-fos mRNA accumulates two times after irradiation of mouse fibroblasts with UV-C. BALB/c 3T3 cells were irradiated with UV (254 nm, 30 J/m^2), treated with TPA (60 ng/ml), or not treated, and the RNAs were analyzed for the amount of c-fos, actin, c-myc, and ornithine decarboxylase (ODC) specific mRNAs as described in Fig. 1. The c-fos and actin have been probed on one blot (after dehybridization), and c-myc and ODC (with another aliquot of the same RNAs) on a second blot.

however, cells can react to UV but lack part of the phorbol ester response. The F9 teratocarcinoma stem cells show this phenotype, although it is yet unknown what they lack. In F9 cells the c-jun gene is not inducible by phorbol esters (Yang-Yen et al., 1990). This is not due to general unresponsiveness of the c-jun gene, because c-jun transcription is inducible by retinoic acid (Yang-Yen et al., 1990) and by UV irradiation (Rahmsdorf et al., in preparation). Also, F9 cells possess all that is needed for c-fos gene activation by phorbol esters (Yang-Yen et al., 1990; König, 1991). The defect must be in a branch of the signaling that is c-jun specific and that differs between UV and TPA induction. As a consequence, UV induces AP-1 binding activity although TPA does not, probably because there is no new c-Jun synthesis and very low constitutive levels (Fig. 3) (Rahmsdorf et al., in preparation).

In HeLa Cells Transcription from the SV40
Enhancer/Promoter is Only Induced by UV, but Not by TPA

The SV40 enhancer/promoter is another case of divergence just prior to the promoter of a specific gene. The enhancer/promoter is composed of two consecutive 72-bp repeats, three 21-bp repeats, and the TATA box (Zenke et al., 1986) (Fig. 4). This sequence linked to a reporter gene and transiently introduced in HeLa cells is strongly activated upon UV treatment of these cells, but not upon TPA treatment (Fig. 4, pSV2CAT; data in collaboration with Pierre Chambon).

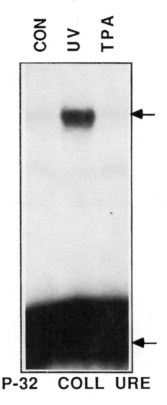

P-32   COLL URE

FIGURE 3. Ultraviolet irradiation, but not TPA, induces AP-1 binding activity in F9 teratocarcinoma stem cells. The F9 teratocarcinoma stem cells were untreated, UV-irradiated (254 nm, 30 J/m^2), or treated with TPA (60 mg/ml). At 4 h after treatment nuclear extracts were prepared and probed for binding to a radioactive double stranded oligonucleotide encompassing the TPA- and UV-responsive element of the human collagenase gene according to Stein et al. (1989a). The upper arrow points to oligonucleotide complexed with protein, and the lower arrow to the free oligonucleotide.

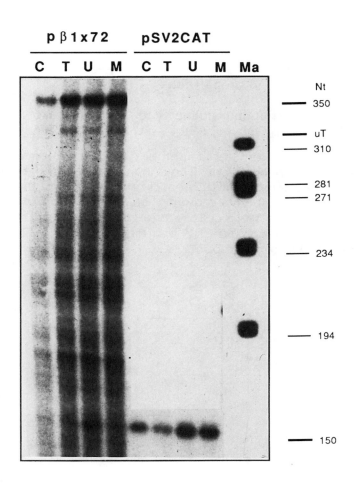

pβ1x72   pSV2CAT

C T U M   C T U M   Ma

Nt

— 350

uT

— 310

— 281
— 271

— 234

— 194

— 150

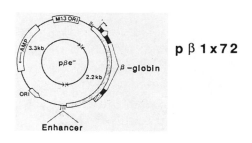

pβ1x72

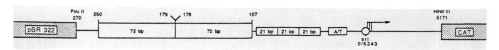

pSV2CAT

In contrast, a gene construct that contains only the distal SV40 enhancer alone behind the β-globin gene (driven by its own promoter) is strongly activated by UV and by TPA (Fig. 4, pβ1×72). Also, subsections of the enhancer react equally well to phorbol esters and to UV. Although not proven, it is likely that nonresponsiveness of pSV2CAT to TPA is due to the presence of the 21-bp repeats: The SV40 AP-1 binding site in front of the three 21-bp repeats and the SV40 promoter could only be induced by UV, while the same site cloned in front of the herpes simplex thymidine kinase promoter was induced by UV and by TPA (M. Vanetti, unpublished, in collaboration with P. Chambon and C. Jonat).

Concluding these previous sections, the data suggest rather substantial differences in the signaling pathways between UV and TPA.

In Co631 Hamster Cells only UV, but Not TPA, Leads to the Amplification of Integrated SV40 Sequences and to the Enhanced Binding of Proteins to the Early Domain

As already discussed, both UV and TPA promote DNA synthesis and induce the amplification of specific genes (Varshawski, 1981; Tlsty et al., 1984). A cell in which gene amplification can be studied without clonal selection (contrary to the case of the amplification of the dihydrofolate reductase gene mentioned earlier) is the Co631 cell, which carries several stably integrated copies of SV40. Upon treatment of these cells with chemical carcinogens or with UV, the integrated SV40 copies were amplified up to 25-fold within the next 72 h (Lavi, 1981, 1982; chapter by Lücke-Huhle, this volume). The TPA treatment of cells, however, did not lead to the amplification of SV40 sequences and did not enhance carcinogen-induced SV40 amplification (C. Lücke-Huhle, unpublished). The carcinogen-induced amplification occurs also in the presence of protein synthesis inhibitors and depends critically on a protein binding domain of the SV40 enhancer/promoter, the so-called early domain (Lücke-Huhle et al., 1989) (Fig. 5a). Destruction of this early domain inhibits SV40 replication in vivo and in vitro (Li et al., 1986), and introduction of a high copy number of the early domain sequence into Co631 cells inhibits UV-

---

FIGURE 4. In contrast to the isolated enhancer, the complete SV40 enhancer/promoter is induced in HeLa cells only by DNA-damaging agents but not by TPA. Logarithmically growing HeLa tk⁻ cells were transfected with 6 μg of either pβ1×72 or pSV2CAT. The pSV2CAT contains the two 72-bp enhancers, the three 21-bp repeats, and sequences up to position 5171 linked to the CAT gene (Gorman et al., 1982). The pβ1×72 contains only the distal enhancer (position 270 to 179) cloned 3' from the β-globin gene, which is driven by its own promoter (Ondek et al., 1987). Transfection was with the DEAE dextran method (Kawai and Nishizawa, 1984) and chloroquine treatment was for 8 h. Thereafter, the cells were starved in DMEM (0.5% FCS) for 24 h and then treated with TPA (T, 70 ng/ml), UV (U, 20 J/m²), or mitomycin C (M, 1 μg/ml). At 24 h later total RNA was prepared and 20 μg were assayed for β-globin transcripts with a radioactive globin RNA (cells transfected with pβ1×72) and for CAT transcripts with a radioactive CAT RNA. Indicative fragments are 350 (β-globin) and 150 (CAT) nucleotides long; uT incorrectly initiated transcripts of pβ1×72.

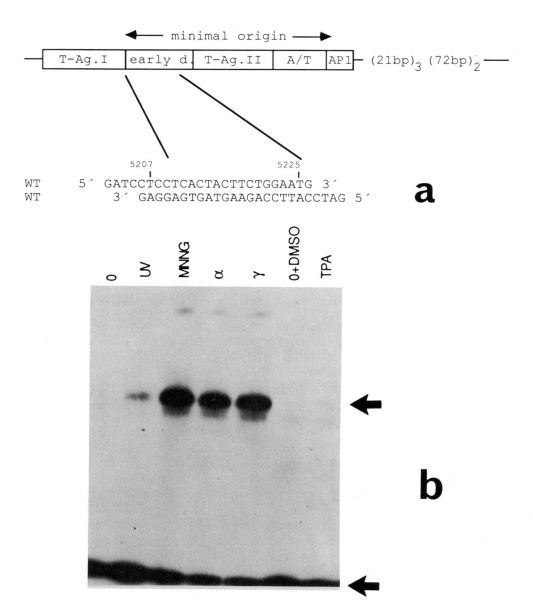

FIGURE 5. DNA-damaging agents induce, in contrast to TPA, enhanced protein binding to the SV40 early domain. (a) The organization of the minimal origin of replication of SV40 is shown together with the SV40 72-bp enhancers and the 21-bp repeats. AP-1, consensus AP1 binding site; A/T, TATA box; T-Ag.II, T-antigen binding site II; early d, early domain; T-Ag.I, T-antigen binding site I. Below is shown the oligonucleotide encompassing the early domain, which was used for the gel retardation experiment shown in (b). Co631 Chinese hamster cells transformed with SV40 (Lavi, 1981) were treated with UV (5 J/m²), N-methyl-N-nitro-N-nitrosoguanidine (MNNG, 10 μg/ml), α-irradiation (5 Gy), γ-irradiation (12 Gy), dimethyl sulfoxide (DMSO, 0.03%) (the solvent for TPA), and TPA (60 ng/ml). After 30 min whole-cell extracts were prepared and 5 μg of protein was incubated with the radiolabeled early-domain oligonucleotide at 25°C for 30 min. Protein-bound oligonucleotide (upper band) was separated from free oligonucleotide (lower band) by electrophoresis (Lücke-Huhle et al., 1989). The experiment shown has been performed by Sabine Mai during her thesis work in Karlsruhe.

induced SV40 amplification (Lücke-Huhle et al., 1989). This suggests that this sequence binds one or more essential factors. Traut and Fanning (1988) have presented evidence for the binding of competence factors at this site. Our own attempts have yielded DNA damage-induced protein binding to the early domain (Lücke-Huhle et al., 1989) (Fig. 5b), which could not be induced by TPA. Yet the identity and function of this protein are totally unclear. Particularly we have not been able to relate the in vitro complex to the in vivo function by point mutations.

Other laboratories have also detected differences between UV- and TPA-induced genetic processes. The DDIA class I and class II genes are inducible by UV, but not by TPA (Holbrook and Fornace, 1991). Without claiming exhaustion of this list, we turn now to the question of how to explain such differences between UV and TPA.

DIFFERENCES IN THE TPA- AND UV-INDUCED SIGNAL CHAINS

Different Primary Targets

The only primary target that has been found for TPA is the family of cytoplasmic serine and threonine specific protein kinases that depend on Ca^{2+} and phospholipids (subspecies α, β_I, β_{II} and γ). Upon binding of TPA, the enzymes are translocated into the plasma membrane and their enzymatic activity is induced. Other protein kinase C (PKC) related enzymes (ϵ, δ, ζ, η) lack one of the conserved regions (C2, which is presumably involved in Ca binding), and PKC η has been shown to reside and to be activated by phorbol esters in the nucleus (Greif et al., 1992; for review see Nishizuka, 1988). The signal flow from the activation of protein kinase C to the responding transcription factors is just being unraveled. The PKC activation leads to the activation of mitogen-activated protein (MAP) kinases, possibly through the action of Ras (Nori et al., 1992); MAP kinases have been shown to phosphorylate c-Jun protein in its N-terminal domain in a way that is indistinguishable from its in vivo phosphorylation after phorbol ester treatment of cells (Pulverer et al., 1991). It is not yet known whether other kinases (such as Raf-1, Kolch et al., 1991; casein kinase II, Gauthier-Rouvière et al., 1991) are involved in this or other phorbol ester-induced signal chains.

We and others have provided evidence that the primary target of UV that starts the signaling process is cellular DNA. It was shown, for instance, that wavelengths in the ultraviolet part of the spectrum between 260 and 280 nm that are best absorbed by DNA are most efficient in inducing the transcription of the c-fos, the collagenase, and the HIV-1 gene (Stein et al., 1989a). The strength of the inducing signal dropped toward longer wavelengths. Ultraviolet radiation reaching the human body surface naturally suffices to induce gene transcription: A 2-h noon-time exposure of cells at 2000 m above sea level on a sunny day inflicts the same amount of DNA damage as is induced by 30 J/m^2 monochromatic UV-C (254 nm, Klocker et al., 1984). These are the doses that in our hands are most efficient in the induction of gene transcription. Another argument for the involvement of DNA in UV-induced gene transcription came from the comparison of the dose dependence of induced gene transcription in DNA repair-proficient (normal) and DNA repair-deficient (xeroderma pigmentosum group A) cells (Miskin and Ben Ishai, 1981; Schorpp et al., 1984; Stein et al., 1989a). These cells differ only in their capability to remove UV-induced thymidine dimers and 6-4-photoproducts. We found that in xeroderma pigmentosum cells 10-fold lower doses were required to induce maximal transcription of collagenase, metallothionein, and HIV-1 than in wild type cells. The nonrepaired UV-induced DNA damage must, therefore, be part of the UV-induced signal chain, and DNA repair prevents signaling. Because initially DNA damage is identical in repair-proficient and repair-deficient cells and differences in damage density can only be expected after hours (of repair in the wild type cells), the induction of the immediate early gene *c-fos* (and certainly of other immediate early genes) requires the same UV dose in wild type and xeroderma pigmentosum cells. The signal to these genes is delivered within minutes after irradiation (Büscher et al., 1988; König et al., 1989).

These findings thus suggest that the cellular DNA is the target of UV irradiation and that it is the nonrepaired DNA damage that transmits the signal. How should DNA damage elicit a signal? Signaling could be induced by a stop of the transcriptional or the replicating machinery at the sites of DNA damage. That arrested transcriptional complexes elicit a signal remains a possibility. The effective damage density can be calculated from dimer determinations: In xeroderma pigmentosum cells 2 J/m^2 is highly efficient; this UV dose causes one pyrimidine dimer in 20,000 bases (Wulff, 1963). Because each dimer potentially arrests a significant number of transcriptional complexes, arrested transcriptional complexes could be involved in signaling. A stop of replication does not seem essential because the UV induction of *c-fos* and collagenase transcription was similarly efficient in growing and G$_0$ arrested primary human fibroblasts (Stein et al., 1989a). A third possibility, yet undocumented, should be mentioned: DNA damage could be recognized by damage-specific or general binding proteins that cause signal transduction.

Protein Kinase C is Involved in TPA- but Not UV-C-Induced Gene Transcription

The bulk of protein kinase C is the target for TPA. In view of the similar characteristics of TPA- and UC-C-induced signaling, we investigated whether protein kinases of this family are necessary intermediates in UV-induced gene expression. The available evidence suggests that they are not: (1) In contrast to TPA, UV-C irradiation of cells does not induce the translocation of protein kinase C from the cytoplasm to the membrane and (2) in cells that have been depleted of protein kinase C by pretreatment with TPA, UV-induced gene transcription still works (Büscher et al., 1988). In view of the increasing number of different protein kinase C species with different properties, however, some of which are localized in the cell nucleus and not down-regulated upon phorbol ester treatment (Greif et al., 1992), the statement that protein kinase C is not involved in UV-induced signaling remains speculative, as long as the protein kinases involved in UV signaling have not been identified. It should be mentioned here that with other types of radiation, namely, long-wavelength UV (UV-A) and x-rays, protein kinase C does seem to be involved in signal transduction. The UV-A leads to the translocation of protein kinase C from the cytoplasm to the cellular membrane (Matsui and De Leo, 1990), and x-ray-induced transcription of the *egr1* and the *c-jun* gene is inhibited in TPA-pretreated cells (Hallahan et al., 1991).

Different Exhaustible Components in TPA- and UV-Treated Cells

We discussed earlier that both UV and TPA induce a refractory cellular state, in which the *c-fos* gene (and the *c-jun* gene) cannot be induced a second time by the same agent. This refractory state could be caused, for instance, by a component of the signal chain that is inactivated or degraded during the process and that is not resynthesized. After its TPA-induced translocation to the cellular membrane, protein kinase C is proteolytically degraded. Because, however, the inactivation of the enzyme proceeds with a half-life of 4 h (Rodriguez-Pena and Rozengurt, 1984), whereas the TPA-induced refractory state is already fully established at 3 h after treatment (Büscher et al., 1988) and possibly much earlier (our unpublished observations), it is rather improbable that the refractory state is due to the proteolytic degradation of protein kinase C. It may be due to the translocation of protein kinase C to the membrane, where it may be not available for a second TPA pulse, or to the inactivation of other members of the signal chain. Experiments in which TPA responsiveness of cells could be restored by the injection of protein kinase C at late times (3 days) after TPA pretreatment suggested that this late refractory state may be due to deprivation of cytoplasmic protein kinase C (Pasti et al., 1986). Similar experiments performed at early time points after TPA treatment should show which of the alternatives mentioned above is correct.

The refractory state induced by UV-C irradiation differs from the one induced by TPA and must be caused by a different exhaustible component. This follows from the fact that in cells that had been pretreated with TPA and that are refractory to this drug, the early genes can be induced by UV, and vice versa (Büscher et al., 1988). There is as yet no information on which of the components of the UV-induced signal chain causes refractoriness.

In Contrast to TPA-Induced Signaling, UV-Induced Signal Transduction is Inhibited by Suramin

We and others have found previously that the culture medium of UV-treated cells contains a factor (or a group of factors) that transduce the UV signal to nonirradiated cells, thereby amplifying the UV response (Schorpp et al., 1984; Rotem et al., 1987; Maher et al., 1988; Stein et al., 1989c). We recently identified the UV-induced active fraction secreted from HeLa cells as the two growth factors interleukin 1α and basic fibroblast growth factor (M. Krämer, C. Sachsenmaier, P. Herrlich, H. J. Rahmsdorf, submitted). In an attempt to determine whether these factors not only transmit the UV-induced signal to nonirradiated cells but also act back on the secreting cell, we attempted to block the growth factor–growth factor receptor interaction. Suramin is a polycyclic compound known to inhibit the interaction of various growth factors with their receptors (Betsholtz et al., 1986; Coffey et al., 1987; Huang and Huang, 1988). In fact, suramin efficiently inhibited gene expression induced by the culture medium from UV-irradiated cells. The expectation was that suramin would block the minor portion of the UV response that would be mediated by growth factors. The outcome totally exceeded the expectation. Suramin also inhibited completely any UV-induced gene transcription in the irradiated cells, while only marginally interfering with TPA-induced gene transcription. With the experimental conditions used, every cell should have received DNA damage, and because the inhibition was complete, growth factor–growth factor receptor interactions are indeed part of the UV-induced signal chain (provided suramin has no yet unknown other effects). Experiments with inactivating antibodies directed against growth factors or growth factor receptors, with anti-sense RNAs inhibiting the synthesis of growth factors or recombinational elimination of the growth factor genes, are needed to confirm the hypothesis.

The picture that we draw here for UV-induced signaling looks very complicated: A signal chain that is initiated by DNA damage and that ends with the activation of transcription factors passes through the cytoplasm and the extracellular space. That UV-induced DNA damage can indeed affect membrane metabolism has been shown recently: UV-C-induced release of arachidonic acid from membrane lipids could be inhibited by treating the cells with longwave UV, a process which, by activating the photolyase, leads to the destruction of UV-induced thymidine dimers (Kaleta et al., 1991).

TPA and UV Induce Different Modifications
of the Transcription Factor AP-1

As discussed earlier, the ultimate targets of the TPA- and UV-induced signal chains are nuclear transcription factors, which are activated by posttranslational modification. It is by now clear that there are major differences in UV- and TPA-induced signal transduction. A consequence would be that the seemingly identical activation of transcription factors is in fact not identical in mechanism. In order to investigate whether the UV- and TPA-induced signal chains meet only at the responding transcription factor or already upstream, we identified the posttranslational modifications provoked by these two inducers. The transcription factor AP-1 should serve as example: both subunits of this heterodimeric molecule, Fos and Jun, are differently modified by the inducers. TPA, in contrast to UV, transforms Fos in 5 min into a species with a slower electrophoretic mobility (Fig. 6; compare control and TPA lines in [^{32}P]- and

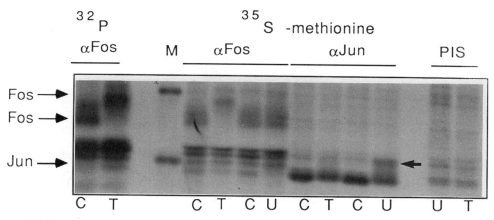

Different modification of AP-1 by phorbol esters and UV

FIGURE 6. Different modification of AP-1 by phorbol esters and UV. 3.5×10^5 HeLa tk$^-$ cells were incubated in phosphate-free DMEM medium for 90 min and then labeled with 18.5×10^6 Bq ^{32}P-orthophosphate per milliliter culture medium for 2.5 h (^{32}P). For labeling with [^{35}S]methionine, 3.5×10^5 HeLa tk$^-$ cells were incubated in methionine-free culture medium with 4.6×10^6 Bq [^{35}S]methionine/ml medium for 3 h (^{35}S-methionine). Thereafter cells were not treated (C), treated with TPA (80 ng/ml), or irradiated with UV (30 J/m^2) and incubated for 30 min. The cells were disrupted with detergents and specific proteins precipitated by antibodies to c-Jun (antibody 891 raised in rabbits against the amino acids 47–59 of c-Jun, Angel et al., 1988; donated by Dr. A. Darling, Glasgow; α Jun) and to c-Fos (antibody X4B, raised in rabbits against a β-gal-Fos-fusion protein containing fos amino acids 52–252 of FBR-fos, and amino acids 275–316 of FBJ-fos; van Beveren et al., 1983, 1984; α Fos). Cell lysates were precleared with a preimmune serum (PIS). Arrows at the left site point to the two major Fos modifications and to the major c-Jun band; the arrow in the figure points to the c-Jun modification, which is specifically induced by UV. The marker line shows two molecular weight standards (46 and 69 kD).

[^{35}S]methionine-labeled cells). Both proteins are phosphoproteins (Fig. 6, the two left lines), but the TPA-induced change in mobility of c-Fos is not due to the differential phosphorylation of the protein. On the other hand, UV treatment of cells, but not TPA, generates in several minutes a Jun species with changed mobility (compare the Jun immunprecipitates from [^{35}S]methionine-labeled cells). Besides, both TPA and UV induce dephosphorylation of the c-Jun protein in the DNA binding domain and enhanced phosphorylation in the transactivating part of the protein. These experiments suggest that UV and TPA may activate a common modifying enzyme, which modifies AP-1 identically in response to both inducers; moreover, both inducers also activate different enzymes, which modify AP-1 differently. It will be important to find out which of the modifications mentioned are functionally relevant.

If UV, as suggested earlier, indeed works through a growth-factor loop, then UV- and growth-factor-induced modifications should be similar. This is indeed what we have found: Interleukin 1α induces a modification of the c-Jun protein that is indistinguishable from the UV-induced c-Jun modification.

CONCLUSIONS

Although reportedly UV and TPA appear to differ in their capacity to induce transformation or tumor promotion, the available evidence does not yield much of an explanation for the putative difference. Both carcinogens, UV and TPA, induce amazingly similar programs of gene expression. Among the genes induced are a number of potentially transforming genes such as c-fos and c-jun. The primary targets of TPA and UV are, however, different: TPA activates PKC directly, but UV needs to be absorbed by DNA. The signal chains starting from the primary targets are consequently different. What is amazing is the comparable outcome.

REFERENCES

Angel, P., Baumann, I., Stein, B., Delius, H., Rahmsdorf, H. J., and Herrlich, P. 1987a. 12-O-tetradecanoyl-phorbol-13-acetate induction of the human collagenase gene is mediated by an inducible enhancer element located in the 5′-flanking region. *Mol. Cell. Biol.* 7:2256–2266.

Angel, P., Imagawa, M., Chiu, R., Stein, B., Imbra, R. J., Rahmsdorf, H. J., Jonat, C., Herrlich, P., and Karin, M. 1987b. Phorbol ester-inducible genes contain a common cis element recognized by a TPA-modulated trans-acting factor. *Cell* 49:729–739.

Angel, P., Allegretto, E. A., Okino, S. T., Hattori, K., Boyle, W. J., Hunter, T., and Karin, M. 1988. Oncogene jun encodes a sequence-specific trans-activator similar to AP-1. *Nature* 332:166–171.

Baeuerle, P. A., and Baltimore, D. 1988a. Activation of DNA-binding activity in an apparently cytoplasmic precursor of the NF-κB transcription factor. *Cell* 53:211–217.

Baeuerle, P. A., and Baltimore, D. 1988b. IκB: A specific inhibitor of the NF-κB transcription factor. *Science* 242:540–546.

Betsholtz, C., Johnsson, A., Heldin, C.-H., and Westermark, B. 1986. Efficient reversion of simian sarcoma virus-transformation and inhibition of growth factor-induced mitogenesis by suramin. *Proc. Natl. Acad. Sci. USA* 83:6440–6444.

Büscher, M., Rahmsdorf, H. J., Litfin, M., Karin, M., and Herrlich, P. 1988. Activation of the c-fos gene by UV and phorbol ester: Different signal transduction pathways converge to the same enhancer element. *Oncogene* 3:301–311.

Coffey, R. J., Jr., Leof, E. B., Shipley, G. D., and Moses, H. L. 1987. Suramin inhibition of growth factor receptor binding and mitogenicity in AKR-2B cells. *J. Cell. Physiol.* 132:143–148.

Cohn, S. M., Krawisz, B. R., Dresler, S. L., and Lieberman, M. W. 1984. Induction of replicative DNA synthesis in quiescent human fibroblasts by DNA damaging agents. *Proc. Natl. Acad. Sci. USA* 81:4828–4832.

Devary, Y., Gottlieb, R. A., Lau, L. F., and Karin, M. 1991. Rapid and preferential activation of the c-jun gene during the mammalian UV response. *Mol. Cell. Biol.* 11:2804–2811.

Diamond, L. 1987. Tumor promoters and cell transformation. In *Mechanisms of Cellular Transformation by Carcinogenic Agents*, eds. D. Grunberger and S. Goff, pp. 73–133. New York: Pergamon Press.

Driedger, P. E., and Blumberg, P. M. 1977. The effect of phorbol diesters on chicken embryo fibroblasts. *Cancer Res.* 37:3257–3265.

Evans, R. M. 1988. The steroid and thyroid hormone receptor superfamily. *Science* 240:889–895.

Gauthier-Rouvière, C., Basset, M., Blanchard, J.-M., Cavadore, J.-C., Fernandez, A., and Lamb, N. J. C. 1991. Casein kinase II induces c-fos expression via the serum response element pathway and p67^{SRF} phosphorylation in living fibroblasts. *EMBO J.* 10:2921–2930.

Gorman, C. M., Moffat, L. F., and Howard, B. H. 1982. Recombinant genomes which express chloramphenicol acetyltransferase in mammalian cells. *Mol. Cell. Biol.* 2:1044–1051.

Greif, H., Ben-Chaim, J., Shimon, T., Bechor, E., Eldar, H., and Livneh, E. 1992. The protein kinase C-related PKC-L(η) gene product is localized in the cell nucleus. *Mol. Cell. Biol.* 12:1304–1311.

Hallahan, D. E., Sukhatme, V. P., Sherman, M. L., Virudachalam, S., Kufe, D., and Weichselbaum, R. R. 1991. Protein kinase C mediates x-ray inducibility of nuclear signal transducers EGR1 and JUN. *Proc. Natl. Acad. Sci. USA* 88:2156–2160.

Härtig, E., Loncarevic, I. F., Büscher, M., Herrlich, P., and Rahmsdorf, H. J. 1991. A new cAMP response element in the transcribed region of the human c-fos gene. *Nucleic Acids Res.* 19:4153–4159.

Herrlich, P., Ponta, H., and Rahmsdorf, H. J. 1992. DNA damage-induced gene expression: Signal transduction and relation to growth factor signaling. *Rev. of Physiol. Biochem. Pharmacol.* 119:187–223.

Holbrook, N. J., and Fornace, A. J., Jr. 1991. Response to adversity: Molecular control of gene activation following genotoxic stress. *New Biol.* 3:825–833.

Huang, S. S., and Huang, J. S. 1988. Rapid turnover of the platelet-derived growth factor receptor in sis-transformed cells and reversal by suramin. *J. Biol. Chem.* 263:12608–12618.

Jonat, C., Rahmsdorf, H. J., Park, K.-K., Cato, A. C. B., Gebel, S., Ponta, H., and Herrlich, P. 1990. Antitumor promotion and antiinflammation: Down-modulation of AP-1 (Fos/Jun) activity by glucocorticoid hormone. *Cell* 62:1189–1204.

Jonat, C., Stein, B., Ponta, H., Herrlich, P., and Rahmsdorf, H. J. 1992. Positive and negative regulations of collagenase gene expression. *Matrix* Supplement No. 1, 145–155.

Kaleta, E. W., Applegate, L. A., and Ley, R. D. 1991. Photoreactivation of ultraviolet radiation-induced release of arachidonic acid from marsupial cells. *Photochem. Photobiol.* 54:747–752.

Karin, M., Haslinger, A., Holtgreve, H., Richards, R. I., Krauter, P., Westphal, H. M., and Beato, M. 1984. Characterization of DNA sequences through which cadmium and glucocorticoid hormones induce human metallothionein-II$_A$ gene. *Nature* 308:513–519.

Karin, M., Haslinger, A., Heguy, A., Dietlin, T., and Cooke, T. 1987. Metal-responsive elements act as positive modulators of human metallothionein-II$_A$ enhancer activity. *Mol. Cell. Biol.* 7:606–613.

Kawai, S., and Nishizawa, M. 1984. New procedure for DNA transfection with polycation and dimethyl sulfoxide. *Mol. Cell. Biol.* 4:1172–1174.

Kennedy, A. R., Mondal, S., Heidelberger, C., and Little, J. B. 1978. Enhancement of X-ray transformation by 12-O-tetradecanoyl-phorbol-13-acetate in a cloned line of C3H mouse embryo cells. *Cancer Res.* 38:439–443.

Kennedy, A. R., Fox, M., Murphy, G., and Little, J. B. 1980. Relationship between x-ray exposure and malignant transformation in C3H 10T1/2 cells. *Proc. Natl. Acad. Sci. USA* 77:7262–7266.

Klocker, H., Auer, B., Burtscher, H. J., Hirsch-Kauffmann, M., and Schweiger, M. 1984. A synthetic hapten for induction of thymine-dimer-specific antibodies. *Eur. J. Biochem.* 142:313–316.

Kolch, W., Heidecker, G., Lloyd, P., and Rapp, U. R. 1991. Raf-1 protein kinase is required for growth of induced NIH/3T3 cells. *Nature* 349:426–428.

Kovary, K., and Bravo, R. 1991. Expression of different Jun and Fos proteins during the G_0-to-G_1 transition in mouse fibroblasts: In vitro and in vivo associations. *Mol. Cell. Biol.* 11:2451–2459.

König, H., Ponta, H., Rahmsdorf, U., Büscher, M., Schönthal, A., Rahmsdorf, H. J., and Herrlich, P. 1989. Autoregulation of fos: The dyad symmetry element as the major target of repression. *EMBO J.* 8:2559–2566.

König, H. 1991. Cell-type specific multiprotein complex formation over the c-fos serum response element in vivo: Ternary complex formation is not required for the induction of c-fos. *Nucleic Acids Res.* 19:3607–3611.

Krämer, M., Stein, B., Mai, S., Kunz, E., König, H., Loferer, H., Grunicke, H. H., Ponta, H., Herrlich, P., and Rahmsdorf, H. J. 1990. Radiation-induced activation of transcription factors in mammalian cells. *Radiat. Environ. Biophys.* 29:303–313.

Lavi, S. 1981. Carcinogen-mediated amplification of viral DNA sequences in simian virus 40-transformed Chinese hamster embryo cells. *Proc. Natl. Acad. Sci. USA* 78:6144–6148.

Lavi, S. 1982. Carcinogen-mediated amplification of specific DNA sequences. *J. Cell. Biochem.* 18:149–156.

Li, J. J., Peden, K. W. C., Dixon, R. A. F., and Kelly, T. 1986. Functional organization of the simian virus 40 origin of DNA replication. *Mol. Cell. Biol.* 6:1117–1128.

Lücke-Huhle, C., Mai, S., and Herrlich, P. 1989. UV induced early-domain binding factor as the limiting component of simian virus 40 DNA amplification in rodent cells. *Mol. Cell. Biol.* 9:4812–4818.

Maher, V. M., Sato, K., Kateley-Kohler, S., Thomas, H., Michaud, S., McCormick, J. J., Kraemer, M., Rahmsdorf, H. J., and Herrlich, P. 1988. Evidence of inducible error-prone mechanisms in diploid human fibroblasts. In *DNA Replication and Mutagenesis,* eds. R. E. Moses and W. C. Summers, pp. 465–471. Washington, D.C.: American Society of Microbiology.

Mallick, U., Rahmsdorf, H. J., Yamamoto, N., Ponta, H., Wegner, R.-D., and Herrlich, P. 1982. 12-O-tetradecanoylphorbol-13-acetate-inducible proteins are synthesized at an increased rate in Bloom syndrome fibroblasts. *Proc. Natl. Acad. Sci. USA* 79:7886–7890.

Matsui, M. S., and De Leo, V. A. 1990. Induction of protein kinase C activity by ultraviolet radiation. *Carcinogenesis* 11:229–234.

Miskin, R., and Ben-Ishai, R. 1981. Induction of plasminogen activator by UV light in normal and xeroderma pigmentosum fibroblasts. *Proc. Natl. Acad. Sci. USA* 78:6236–6240.

Nishizuka, Y. 1984. The role of protein kinase C in cell surface signal transduction and tumour promotion. *Nature* 308:693–698.

Nishizuka, Y. 1988. The molecular heterogeneity of protein kinase C and its implications for cellular regulation. *Nature* 334:661–665.

Nori, M., L'Allemain, G., and Weber, M. J. 1992. Regulation of tetradecanoyl phorbol acetate-induced responses in NIH 3T3 cells by GAP, the GTPase-activating protein associated with $p21^{c-ras}$. *Mol. Cell. Biol.* 12:936–945.

Ondek, B., Shepard, A., and Herr, W. 1987. Discrete elements within the SV40 enhancer region display different cell-specific enhancer activities. *EMBO J.* 6:1017–1025.

Pasti, G., Lacal, J.-C., Warren, B. S., Aaronson, S. A., and Blumberg, P. M. 1986. Loss of mouse fibroblast cell response to phorbol esters restored by microinjected protein kinase C. *Nature* 324:375–377.

Pulverer, B. J., Kyriakis, J. M., Avruch, J., Nikolakaki, E., and Woodgett, J. R. 1991. Phosphorylation of c-jun mediated by MAP kinases. *Nature* 353:670–674.

Rahmsdorf, H. J., Schönthal, A., Angel, P., Litfin, M., Rüther, U., and Herrlich, P. 1987. Posttranscriptional regulation of c-fos mRNA expression. *Nucleic Acids Res.* 15:1643–1659.

Rahmsdorf, H. J., and Herrlich, P. 1990. Regulation of gene expression by tumor promoters. *Pharm. Ther.* 48:157–188.

Rodriguez-Pena, A., and Rozengurt, E. 1984. Disappearance of Ca^{2+}-sensitive, phospholipid-dependent protein kinase activity in phorbol ester-treated 3T3 cells. *Biochem. Biophys. Res. Commun.* 120:1053–1059.

Rotem, N., Axelrod, J. H., and Miskin, R. 1987. Induction of urokinase-type plasminogen activator by UV light in human fetal fibroblasts is mediated through a UV-induced secreted protein. *Mol. Cell. Biol.* 7:622–631.

Schorpp, M., Mallick, U., Rahmsdorf, H. J., and Herrlich, P. 1984. UV-induced extracellular factor from human fibroblasts communicates the UV response to nonirradiated cells. *Cell* 37:861–868.

Sorger, P. K. 1991. Heat shock factor and the heat shock response. *Cell* 65:363–366.

Stein, B., Rahmsdorf, H. J., Steffen, A., Litfin, M., and Herrlich, P. 1989a. UV-induced DNA damage is an intermediate step in UV-induced expression of human immunodeficiency virus type 1, collagenase, c-fos, and metallothionein. *Mol. Cell. Biol.* 9:5169–5181.

Stein, B., Gebel, S., Rahmsdorf, H. J., Herrlich, P., and Ponta, H. 1989b. Different proteins bind to the phorbol ester responsive sequences in collagenase and c-jun promoters. In *Gene Regulation and AIDS: Transcriptional Activation, Retroviruses, and Pathogenesis*, Advances in Applied Biotechnology Series, vol. 7, ed. T. S. Papas, pp. 37–43. The Woodlands, Tex.: Portfolio.

Stein, B., Krämer, M., Rahmsdorf, H. J., Ponta, H., and Herrlich, P. 1989c. UV-induced transcription from the human immunodeficiency virus type 1 (HIV-1) long terminal repeat and UV-induced secretion of an extracellular factor that induces HIV-1 transcription in nonirradiated cells. *J. Virol.* 63:4540–4544.

Stein, B., Angel, P., van Dam, H., Ponta, H., Herrlich, P., van der Eb, A., and Rahmsdorf, H. J. 1992. UV-induced c-jun gene transcription: Two AP-1 like binding sites mediate the response. *Photochem. Photobiol.* 55:409–415.

Tlsty, T. D., Brown, P. C., and Schimke, R. T. 1984. UV radiation facilitates methotrexate resistance and amplification of the dihydrofolate reductase gene in cultured 3T6 mouse cells. *Mol. Cell. Biol.* 4:1050–1056.

Traut, W., and Fanning, E. 1988. Sequence-specific interaction between a cellular DNA-binding protein and the simian virus 40 origin of DNA replication. *Mol. Cell. Biol.* 8:903–911.

Treisman, R. 1985. Transient accumulation of c-fos RNA following serum stimulation requires a conserved 5′ element and c-fos 3′ sequences. *Cell* 42:889–902.

van Beveren, C., van Straaten, F., Curran, T., Müller, R., and Verma, I. M. 1983. Analysis of FBJ-MuSV provirus and c-fos (mouse) gene reveals that viral and cellular fos gene products have different carboxy termini. *Cell* 32:1241–1255.

van Beveren, C., Enami, S., Curran, T., and Verma, I. M. 1984. FBR murine osteosarcoma virus. II. Nucleotide sequence of the provirus reveals that the genome contains sequences acquired from two cellular genes. *Virology* 135:229–243.

Varshawski, A. 1981. Phorbol ester dramatically increases incidence of methotrexate-resistant mouse cells: Possible mechanisms and relevance to tumor promotion. *Cell* 25:561–572.

Whitham, S. E., Murphy, G., Angel, P., Rahmsdorf, H. J., Smith, B. J., Lyons, A., Harris, T. J. R., Reynolds, J. J., Herrlich, P., and Docherty, A. J. P. 1986. Comparison of human stromelysin and collagenase by cloning and sequence analysis. *Biochem. J.* 240:913–916.

Wulff, D. L. 1963. Kinetic of thymine photodimerization in DNA. *Biophys. J.* 3:355–362.

Yang-Yen, H.-F., Chiu, R., and Karin, M. 1990. Elevation of AP1 activity during F9 cell differentiation is due to increased c-jun transcription. *New Biol.* 2:351–361.

Zenke, M., Grundström, T., Matthes, H., Wintzerith, M., Schatz, C., Wildeman, A., and Chambon, P. 1986. Multiple sequence motifs are involved in SV40 enhancer function. *EMBO J.* 5:387–397.

INDEX: GENETIC ELEMENTS, GENE PRODUCTS AND INDUCIBLE FUNCTIONING